12-25-95

Dear Charlotte and Mike,

What a book! I [...]
enjoy it. Merry Christmas!

Love,
Bob and Beverley

EVIDENCE NOT SEEN

EVIDENCE NOT SEEN

A Woman's Miraculous Faith in the Jungles of World War II

Darlene Deibler Rose

A Ruth Graham Dienert Book

HarperSanFrancisco

A Division of HarperCollins*Publishers*

New York, Grand Rapids, Philadelphia, St. Louis
London, Singapore, Sydney, Tokyo, Toronto

EVIDENCE NOT SEEN: *A Woman's Miraculous Faith in the Jungles of World War II*. Copyright © 1988 by Darlene Rose. All rights reserved. Printed in the United States of America. No part of this book may be used or reproduced in any manner whatsoever without written permission except in the case of brief quotations embodied in critical articles and reviews. For information address HarperCollins Publishers, 10 East 53rd Street, New York, NY 10022.

FIRST HARPER & ROW PAPERBACK EDITION PUBLISHED IN 1990.

Library of Congress Cataloging-in-Publication Data

Rose, Darlene Deibler.
    Evidence not seen : a woman's miraculous faith in the jungles of World War II / Darlene Deibler Rose.
        p.   cm.
    "First Harper & Row paperback edition"—T.p. verso.
    "A Ruth Graham Dienert book."
    ISBN 0–06–067020–7 :
    1. Rose, Darlene Deibler.   2. World War, 1939–1945—Prisoners and prisons, Japanese.   3. World War, 1939–1945—Personal narratives, American.   4. Prisoners of war—Indonesia—Celebes—Biography. 5. Prisoners of war—Japan—Biography.   6. Prisoners of war—United States—Biography.   7. Missionaries—New Guinea—Biography. 8. Missionaries—United States—Biography.   9. World War, 1939–1945—Women—Biography.   I. Title.
[D805.N44R67   1990]
940.54′72′52—dc20   90–35975
                                                                          CIP

95   RRD(H) 10 9 8

DEDICATION

An old Roman coin was found on which was a picture of an ox, one of the servants of man. The ox was facing two things, *an altar and a plough*. The inscription read, *"Ready for either."*

To those servants of God who made the supreme sacrifice upon the *altar of martyrdom* in Japanese prisoner of war camps or in the fetid jungles of Borneo:

Rev. C. Russell Deibler   Rev. Andrew Sande
Dr. Robert A. Jaffray   Mrs. Helen Sande
Rev. W. Ernest Presswood   Baby David Jerome Sande
Rev. John F. Willfinger   Rev. Fred C. Jackson

And to those who put their hands to the *plough of service* for the Master, never looking back, surviving the imprisonment, deprivation, separation from and loss of loved ones, hunger, beatings, and rape, and bearing in their bodies the cost of following the Master wherever He led, until He called them home:

Mrs. Robert A. Jaffray   Miss Philoma R. Seely
Miss Margaret M Jaffray   Mrs. Mary A. Dixon
Miss Lilian F. Marsh   Miss Grace M. Dittmar
Miss Margaret E. Kemp

To these, my beloved fellow missionaries, who loved me, guided me, and inspired me, the youngest missionary among them, I lovingly dedicate this book. Out of the past and from the distant shore, I hear them calling, "Run, Darlene, run! Run with perseverance the race that is set before you. *It is worth it all!*"

And so I run!

ACKNOWLEDGMENTS

More than ten years ago I began to write the story of my experiences during World War II for Bruce and Brian, my sons. I wished them to know, if ever difficult circumstances came into their lives, that their mother's God is still alive and very well, and His arm has never lost its ancient power! I owe a special debt of gratitude to them and their father, Jerry, who supported and encouraged me to finish the story.

To those who graciously offered the hospitality of their homes while I worked on the manuscript, I wish to express my thanks: the Kho Hong Gan family of Sydney, Australia; Dr. and Mrs. Robert S. Brown of Omaha, Nebr.; Cornelius and Claire Pals of Golden Valley, Minn.; Betty Knudsen and Ann Dennert of Boone, Iowa; Mrs. Ruth Graham, Montreat, N.C.; and Art and Jackie Holsworth of San Leandro, CA.

I am indebted to those who proofread the manuscript and made helpful suggestions: May Brown and Elizabeth Churchhill of Omaha, Nebr.; Nina Hoffman, Betty Knudsen, Ann Dennert, and Annamae Reed of Boone, Iowa; Elsie David Allen, a fellow prisoner, now of Sydney, Australia, and Molly Farlow of Pasadena, CA.

I was also greatly aided in the typing of the manuscript by Faith Sinclair of The Entrance, NSW Australia; Ruth Cozette of Omaha, Nebr.; and Betty Goodwin of St. Petersburg, Fla.

Finally I want to express sincere appreciation to Karen Mains, who transcribed the tapes I had made and encouraged me to write my own story; to Ruth "Bunny" Graham Dienert, who presented my manuscript to Harper & Row and has continued to encourage and support me; as well as to Roy M. Carlisle and Rebecca Laird, my editors at Harper & Row, whose kindness and assistance have made this book possible.

PREFACE

Peace came to the Japanese prison camp at Kampili two weeks after it had been heralded in the highroads and footpaths, the boulevards and cobbled tracks of the rest of the world. On September 19, 1945, seventeen days after the truce had been signed aboard the battleship USS *Missouri* in Tokyo Bay, I stepped carefully, balancing my emaciated eighty-pound frame, into a bobbing rowboat that was to carry me from Celebes, the island of my captivity, to a flying boat lying at anchor in the harbor sent to evacuate American personnel from the military prison camp.

Eight years before, and a war away, I had arrived in the islands with my husband, on our first wedding anniversary, to begin missionary work in the interior of New Guinea. Now, rowing away from the shore, I could think of nothing but two lonely wooden crosses, half-hidden on some remote hillside. One was the grave of the Reverend C. Russell Deibler, my husband; the other, that of Dr. Robert Alexander Jaffray, my spiritual mentor and an inimitable pioneer and visionary, who had given more than forty years in missionary service to China, Indochina, and Indonesia.

Now alone, I started the journey back to my homeland. How desolate the island shoreline seemed, despite the lush foliage and sparkling blue waters. I turned my face away as great bitterness corroded the edges of my soul like acid. Twenty-eight years old, already a widow for more than two years, I was returning to the United States without a single possession. My mementos and private keepsakes of married life had been pilfered or destroyed. I wore borrowed, ill-fitting clothes. A huge ulcer was eating into the flesh of one leg, and my once soft and fair skin was scarred and mottled from the hours I had spent working in the beastly tropical sun to advance the Japanese war effort. The diseases of imprisonment—beriberi, malaria, and dysentery—had left me frail and debilitated.

For almost four years, my fellow missionaries and I, along with

1,600 other women and children, measured our days in forced labor, meted the hours in separation and deprivation, and marked the anniversaries of the deaths of loved ones who succumbed, one by one, to disease, starvation, and the horrific bombings.

No other world had come to exist beyond the margins of a few barren acres for the hundreds of us incarcerated at Kampili. We were totally isolated; and the cataclysmic events that raged and scourged over the rest of the earth were interpreted to us solely through the eyes and whims of our Japanese captors.

Not a single letter arrived from home. Not one humane Red Cross package or one encouraging pamphlet was ever dropped by the Allies or ferreted under the barbed wire to assure us that someone was fighting for our freedom.

Now, suspended between captivity and a new life, I felt a fear I'd never known. Would I know how to live outside of the confined yet familiar regimen of suffering? Would I ever be free of the recurring, terrifying nightmares of trying to rescue people caught in burning buildings? Would I ever be able to close my eyes without seeing a young Armenian lad lying dead before me on a bamboo bed, one leg missing, cruelly amputated by an incendiary bomb, or stop hearing the ceaseless drip, drip of his body fluids splashing on the ceramic tiles of the camp commander's office floor? Would I ever forget the mute appeal in the eyes of what had been a beautiful blonde woman—a napalm victim of the bombing raid over our camp?

When would the mention of the Kempeitai Secret Police cease to fill me with terror? or the sound of a plane not make me want to run and hide from the incendiary or shrapnel bombs the plane surely carried?

And when would seeing others delirious with joy at being reunited not constrict my heart, still aching with deep pain from my loss of Russell?

Suddenly I was awash in a sea of great bitterness. "Lord, I'll never come back to these islands again. They've robbed me of everything that was most dear to me." The rowboat reached the flying boat bound for intermediate stops in Borneo and the Palawan Island, then on to Manila. "Will there be healing for such hurt?" I could only cry out to God and hope. Healing would have to come if I were ever to truly live—whole and complete again.

Reaching up to grab the rope ladder dangling from the blister of the flying boat, a Catalina, I heard noises from the beach—running feet and calling voices. *Selamat djalan!* "A peaceful journey!" the Indonesian voices rang out across the distance. Those who had come to know the Lord in our Macassar Gospel Tabernacle and those who had shared in the indescribable suffering of imprisonment stood waving in a group. Notification of our departure had come so suddenly that I had been unable to say goodbye.

Their voices were raised in a sweet benediction: "God be with you, till we meet again. . . . " Their song released the waters of bitterness that had flooded my soul, and the hurt began to drain from me as my tears flowed in a steady stream. The healing had begun. I knew then that someday, God only knew when, I would come back to these my people and my island home.

As the Catalina became airborne, carrying me away from the bomb-scarred terrain, the flooded rice fields, the coral coastline, and the mountains of my long bondage, I handed over eight long years of my life into the faithful, wise hands of a gracious God Who alone could help me to understand the mysteries of deep pain and suffering.

With the exception of Darlene Deibler's family and fellow missionaries, certain other names have been changed.

CHAPTER ONE

After six months studying the Dutch language in Holland, C. Russell Deibler, my husband, a veteran missionary, and I, his young bride, set sail aboard the RMS *Volendam,* en route to the Netherlands East Indies.

My first glimpse of the East Indies showed me a veritable hot, humid Eden. The more than 13,500 islands scattered from the South China Sea to the Indian Ocean were bathed in two annual monsoons that left a girdle of mud around most of the larger islands. Each had its own version of swamplands and impregnable rain-forest jungles, with vaulting cathedral-like gloom. Many had active volcanoes that still spewed occasional but deadly flames and lava. All were outlined with coral reefs, quiet lagoons, white-sand beaches fringed with coconut palms, hibiscus, and frangipani bending languidly to the soft ocean breezes. "Ah," I reveled, "an island paradise."

We landed in Batavia, Java, on August 18, 1938, our first wedding anniversary. The scents of my new homeland were foreign yet provocative, exceedingly different from any odors previously known to one accustomed to midwestern Iowa farmland. Each island, each area was different. Some had sulphurous mangrove swamps and decaying jungle flavors. Some reeked of copra. I identified cinnamon, nutmeg, and cloves in the breeze blowing off the Spice Islands. Elsewhere the smell of sea salt was mixed with the heavy perfume of night-blooming jasmine.

Walking through the markets—a patchwork of makeshift tables covered with brightly colored fruits and vegetables, native woven goods, earthenware pots, beautiful sarongs, and trinkets of gold and silver—proved much more invigorating than any American supermarket. The merchants click-clacked two pieces of wood together and called out in sing-song voices the nature of their wares. Nothing had a marked price. I walked away when I first heard a price quoted that was double the value. . . . *Boleh tawar!*

Boleh tawar! the distressed merchants exclaimed, inviting me back to bargain. And bargain I did!

Everything was interesting and intriguing. Immediately I knew a oneness with the people and the place. I pestered Russell with a thousand questions or more.

In the open city canal, happy, chattering men, women, and children were bathing, washing clothes, preparing vegetables, merrily splashing one another, or taking care of bodily functions—all in close proximity to one another.

The train trip to Surabaja through central Java threaded in and out of a mosaic of terraced rice fields and tea plantations.

Three days later, aboard an interisland steamer, Russell and I headed toward Celebes, the mission headquarters in the Netherlands East Indies.

Macassar, the capital and chief seaport of Celebes, was a magnificent tropical city. White-sand beaches extended right up to the terraced rice paddies. A large, very old fort, with an outdated cannon, kept a watchful eye on the harbor. Ocean-going vessels dropped anchor to unload their imported manufactured goods in exchange for a cargo of copra, coffee, rice, corn, salt, and an array of exotic spices.

Russell, never the best of sailors, had recovered sufficiently to join me at the rail after the gangplank had been lowered onto the wharf. He directed my gaze toward the group gathered to the right of the gangplank.

"The tall lady is Margaret Kemp, from Endicott, New York," Russell said. "She and the other single ladies work in the headquarters office and teach in the Bible School."

I recognized Lilian Marsh, for she was the image of her sister, Ethel, a soft-spoken Englishwoman whom I had met in London. They were the daughters of a well-known British minister and author, F. E. Marsh. Both had served many dangerous years in embattled China before Lilian was transferred to the Netherlands East Indies. As I looked at the petite lady, perhaps no more than five feet tall, with her fine, curly hair pulled into a neat bun at the back of her head, I could scarcely believe that she had braved the Boxer Rebellion in Wuchow, as had Philoma Seely, who stood next to her.

Philoma, earlier described to me by Russell as intense and a bit eccentric, was shorter than Lilian. Her gray hair, catching the tropical sun, shone like polished silver. Philoma, tone-deaf, yet miraculously fluent in Chinese, kept the books for the mission headquarters, taught in the Bible School, and was actively involved in the ministry of the local Chinese church.

At the end of the line of single women was Margaret Jaffray, the slightly plump, much-beloved daughter of Dr. and Mrs. Jaffray, the field chairman of the Netherlands East Indies mission. Her dark hair was shot through with silver. Rimless glasses sat atop a pug nose but didn't diminish the humor that sparkled in her hazel eyes.

"Welcome back, stranger!" a gentleman with the group called. They all began waving.

"That's Wesley Brill, head of the Bible School; his wife, Ruby; and their little girl, Donna," Russell explained.

In fear and trepidation I disembarked. The Brills reached Russell first with a hearty greeting, and as I hung back, a little unsure, the two Margarets, Lillian, and Philoma surrounded me. I looked hesitantly into their faces, fearing that one as young as I might not be readily accepted; but they extended kind words of welcome, which I soaked up like a sponge. From that moment I knew a respect and love for them that close proximity in the work, war, and suffering never altered.

The Brills told us that, for the present, Russell and I would live at the mission guest house situated near the edge of the city.

The guest house had large, airy, sparsely furnished bedrooms that opened onto the communal dining and sitting rooms. The cooking area, dipper bath, toilet, and quarters for the hired help were housed in a separate building connected by a covered walkway to the main house. Ceramic tiles made the floor deliciously cool underfoot.

After lunch everyone disappeared—siesta time. With a twinge of guilt, I crawled in under my mosquito net. There was no electric fan, and the heat was oppressive; I slept only to awaken feeling soggy and unrested. I welcomed the dipper bath!

After a refreshing cup of tea, the house emptied, each to his or her own duties. Russell and Wesley went to the shipping office to check on our trunks. I unpacked our suitcases, then sat at the foot of my bed looking out the window, trying to identify all of the new and different sounds and smells, thinking how privileged I was to be there.

Mr. Brill informed me at the evening meal that my language

teacher would be arriving at eight-thirty the following morning. Soon I would be able to communicate with the native people around me beyond a handshake and a smile. God had called me, and He would equip me for the ministry. Language was a tool I would learn to use—and with His enabling, I would use it well.

Promptly on time the next morning I was introduced to a middle-aged Indonesian man who nodded, and I nodded back; then Russell left the room. I spoke not a word of Indonesian; he, not a word of English.

Immediately my tutor rose from his chair and left the room, only to soon reappear. He half-bowed and said, *Selamat pagi, Njonja!*

I stared at the ceremony silently. "Yes, sure," I thought to myself. "Slam it, or whatever that was you said."

When I didn't speak, my tutor again walked out the door and came back in, enunciating every syllable distinctly, *Selamat pagi, Njonja.*

The wheels of my mind began to turn. This was some kind of greeting! So when he walked out the door the third time, I was prepared. He entered, bowed, and repeated the greeting.

I rose from my chair, bowed, and replied, *Selamat pagi, Njonja.*

The short, obese man in a starched, white cotton suit scowled, his hands waving wildly in the air, his head shaking a definite no. He tapped his chest and expressively emphasized, *Tuan, Tuan.* Then he pointed to me and explicitly declared, *Njonja, Njonja.*

I wanted badly to laugh at myself, for I had just replied, "Good morning, Missus," to this intense little man. To assure him that I understood, I bowed and said, *Selamat pagi, Tuan,* to which he replied with a smile, *Baik, baik!* "Good, good!" I didn't know what a "bike" had to do with it, but I learned my first lesson well and never again called a gentleman "Missus."

Every day I studied with my tutor, who proved to be an excellent, if demanding, instructor. When he took his final bow each afternoon, leaving me free time, I gravitated toward the servants' quarters. The cook, a young woman, and the laundry boy helped, laughed at, and corrected my Indonesian. If my pronunciation erred, or I asked them to repeat something, they increased the volume with each repetition. It was with glee that I finally learned to tell them that my hearing was fine; it was my Indonesian that I had to work on!

Celebes was an island of contrast, with a spine of high mountains running north and south that towered to 11,000 feet and mysterious blue lakes that were thousands of feet deep. Lush, tropical vegetation blanketed the mountain terrain, yet near the center of the island, the mountains were honeycombed with limestone caves.

I loved the island and its people. However, I knew it was but a temporary stopover. Soon I would walk over the horizon of the future into the "last unknown"—New Guinea—and I was confident that on this island of Celebes was a group of dedicated, likeminded men and women who would walk the trails with us by faith and prayer.

One September morning I was to meet one of those men, Dr. Robert A. Jaffray.

"That's Mr. Jaffray next to Margaret," Russell explained as the ship docked.

I knew him at first sight. How could I miss him? He stood head and shoulders above his fellow passengers and removed his white pith helmet to acknowledge our greetings. His short, white hair was neatly combed, and he was smooth-shaven except for a closely trimmed mustache. He stood with the natural dignity of a man well born, hands resting on the rail of the ship, one thumb hooked through the handle of a small briefcase.

Margaret—or Muggins or Muggie, as he affectionately called his daughter—helped him ashore and the many packages she carried as gifts for the missionaries adorned her like ornaments on a well-decorated Christmas tree.

Dr. Jaffray approached me and grasped my hand warmly in his. "This must be Darlene. I can understand now why Russell needed a longer furlough." His warm smile and twinkling blue eyes belied any sternness I thought might accompany a man so tall and striking in appearance.

"I'm afraid we're still occupying your house," I apologized, foregoing a greeting in my nervousness.

Margaret hurried to kindly reassure me: "You mustn't think of moving. Mother's visiting friends in Singapore and Java. She'll probably not arrive until after conference."

The field conference for the missionaries in the Netherlands East Indies would not convene until November; that left much time to drink deeply from Mr. Jaffray's contagious enthusiasm for New

Guinea. He carried a well-worn map of the trail from Oeta to the Wissel Lakes with him. As soon as he was settled in, we began to make lists of supplies and cost estimates. The three of us were entirely consumed with the plan of claiming the interior of New Guinea for the Lord. We studied every bit of information available.

We learned that on January 1, 1937, a Dutch pilot named Mr. Wissel and his American copilot, Mr. Jack Atkinson, were flying on an aerial survey job for the Babo Oil Company of New Guinea. The new year had dawned; it was a rare day when the clouds had lifted from New Guinea's backbone, exposing the snow-covered Carstensz Mountain Range. The flyers flew along some thirty miles of snow fields, which often rose to rugged, snow-clad peaks and glacial lakes. They were awed by viewing nature in its infancy— untouched, unspoiled, undesecrated by the needs of modern people. When they dropped over the northern escarpment of the range, they saw below them what appeared to be three circular clouds nestled in the mountains.

Descending to a lower altitude they discovered, not cloud formations, but three crystal-clear lakes, and on the largest of them were men and women rowing canoes in this part of the world said to be uninhabited!

As Russell and I researched the resulting aerial surveys, we learned that the Oeta River, which emptied into the Banda Sea on New Guinea's south coast, had its beginnings in the largest of the Wissel Lakes, Lake Paniai. If someone followed the Oeta River to its source, surely additional villages of the newly discovered people could be found.

The Dutch government, which controlled the East Indies, soon made funds and field police available to make the virgin trip. With the river as their guide, the patrol hacked and tore their way through the dense undergrowth of tropical jungle. They struggled over forest-clad mountains only to find another range awaiting them, steeper, more rugged than the last. Crude lean-tos were haphazardly built at the end of each exhausting day's trek. Many carriers, mostly natives from the coastal regions, died and were laid in shallow graves along the edge of the night's encampment. All suffered from the scant food supply and the freezing temperatures in the higher altitudes. When, nearly a month after leaving the coast, the survivors crossed the last range and walked down into the Stone Age, a great horde of yodeling, near-naked aboriginal people followed them to the headwaters of the Oeta River—Lake Paniai.

A few police, several prisoners, and an assistant patrol officer were left at Enarotali to occupy the hastily constructed government shelters on a hillside overlooking the lake—the only link between this primitive post and the outside world was a battery-powered two-way radio.

The more we heard, the dearer the map and history of New Guinea became. Claimed by Spain in 1545, New Guinea was so named by the Spaniards because the coastal inhabitants of this second-largest island in the world resembled the tribes of Africa's west coast. Dutch traders established the first European outposts on the island early in the nineteenth century, and the Netherlands annexed the western half of the island in 1828.

We came to know every curve on the map of New Guinea. The island resembled a gigantic Paleozoic bird of prey—its primal head thrown up, mouth open, as though to suck in the smaller islands of the Moluccas with the incoming tide. The breast of New Guinea hovered above the topmost tip of Australia, and its tail rested in the Coral Sea. The formidable coastal frontier, the impenetrable mangrove swamps possessively guarded by crocodiles, venomous snakes, and cannibal tribes, had discouraged even the most courageous seafarers for centuries.

On June 21, 1938, an expedition sponsored by the American Museum of Natural History of New York City, led by Richard Archbold, traced the Baliem River through a valley situated 250 miles east of the Wissel Lakes. Their plans were to take samples of the flora and fauna in the foothills of Mount Wilhelmina, one of the highest peaks of New Guinea. Their initial flight into the area revealed a densely inhabited valley of Stone Age cannibals.

Rereading the account of the expedition's findings impressed us anew with the magnitude of the task of reaching the countless millions in New Guinea's mountain recesses.

The field conference convened in November at Benteng Tinggi, in the mountains some sixty kilometers from Macassar. Benteng Tinggi means "High Fort," and it was indeed a place of refuge from the enervating heat of the coast. Meetings were held in a huge octagonal building with adjoining private accommodations.

My first conference was one of much laughter and camaraderie with other missionary families.

The conference voted unanimously that Russell and I, along with another couple, Walter and Viola Post, be appointed to the Wissel Lakes. Even before the conference appointment, we all had obtained work permits, but the government refused to allow women to travel over the rugged and grave-lined trail that already testified to the unbelievable expenditure of strength and endurance demanded of those who traveled inland. Until conditions were favorable, Russell and Walter Post would trek to the Wissel Lakes, and Viola and I would remain in Macassar.

In early December, Walter Post and Russell left by interisland steamer for Ambon, the seat of the government controlling the Spice Islands. Here they purchased additional trail boots from the field police canteen, a minimum of camping equipment, as well as trail food for themselves and their carriers.

The governor was most helpful and arranged for them to travel by government steamer to the village of Oeta on New Guinea's south coast. They arrived safely and unloaded the supplies; then Walter Post returned to Ambon. The day after Christmas, the Ambonese government official took Russell, his ten native carriers, and supplies in the government launch upstream, with a canoe in tow, to the first series of rapids. From this point the supplies and canoe had to be portaged around the rapids. They reached the base camp in late afternoon of the third day. This I learned from a letter forwarded to me through the kindness of the government official.

Remembering what I had read of the trail, I commended Russell and the carriers continually to the Lord for strength and safety, courage and endurance. With the other missionaries, I prayed that God would overshadow him, enabling him to reach the Wissel Lakes speedily and safely. How very much he suffered—how imperative our intercession—I was not to know for many weeks to come.

While Russell was gone, Margaret Kemp pressed me, encouraging me to attempt more than I in myself thought possible. Translating weekly Sunday School lessons for the Junior Department and supervising several teachers—after only a few months of language study—were her idea. I was regularly scheduled to speak in chapel, and the wives of the national workers asked if I would assist them with kindergarten for the children of the teachers and students. It was valuable experience.

I continued furthering my study of the language with a government school teacher from Menado. His excellence in Indonesian inspired me to strive for mastery in the language. Anyone can babble along in Pasar (market) Malay after a few week's study, but proper Indonesian is beautiful, without harsh gutterals. To hear the language well spoken is like listening to a symphony played upon the instrument of words—like enjoying a Renoir painted with the brush of fluency, a masterpiece of lights and shadows.

At the beginning of the new school year, I was handed a Church History book in English and asked if I would accept it as one of the subjects I would be teaching second-year students. I accepted it readily, but after perusing the first few chapters, I began to think I had more enthusiasm than good sense. How do you present Church History to students who, fresh from the jungle, have seen their own coastline, the ocean, and an outside world for the first time within the past year? It was quite an assignment, which proved not only a challenge but an inspiration!

Most of those in my first-year classes were Dyaks from Borneo. Since all classes were in Indonesian, they had to acquire a working knowledge of a language foreign to them. Many of them had never held a pencil before. Besides their language classes, they were taught reading, writing, arithmetic, music, Scripture memorization, and stories of the Old and New Testaments. Being confined to the classroom was in itself a great trial to them; they had been a free people living on the periphery of civilization. Never had they been subjected to a schedule. I loved and admired them as they sat bowed over their books, foreheads creased with frowns; from the extreme heat and intense concentration, trickles of perspiration coursed down their cheeks.

I found fulfillment, joy, and tremendous satisfaction in all my responsibilities. This was one of God's priceless gifts to me—to have had to learn to assume responsibilities from the time I was small. I never had time to be bored! I wouldn't have chosen the separation from Russell, but the Lord had promised me, when I responded to the missionary calling, "Go, . . . I am with you always"! The presence of God and His people kept my heart at peace.

In February word came that Russell would be arriving from

Manokwari. My excitement was without bounds, but why Manokwari, a village on the north coast of New Guinea? When the steamer eased into port at Macassar, I was nearly bursting with excitement. Positioned at the front of those gathered to greet disembarking passengers, I was totally dismayed when I saw a gaunt, wasted stranger with Walter Post.

The other missionaries recognized the slimmed-down Russell as the man they knew before his furlough. But where was the man I had married, the husband who had left for New Guinea? In just eighteen days on the trail and a few months of meager rations, he had lost more than sixty pounds!

"Darlene?" When I heard his voice, I knew it was Russell, but the voice shouldn't belong to this emaciated stranger. I quickly looked down, not wanting him to see my uneasiness. My shyness amused him, but the shock I felt, thinking of what he must have suffered, was with me for days.

He walked with considerable pain, and once back at the house, when he removed his shoes and socks, I knew why. There was no skin on his insteps, the balls of his feet, or any of his toes. He had a serious, advanced case of jungle rot.

Dr. Jaffray immediately sent for a doctor, who came to the house to examine Russell's feet. Turning to me, he said, "Do you see this tissue that is sloughing off? Each morning take a tweezer and tear off every layer until you reach the raw, throbbing flesh. Don't apply the ointment that I'm giving you until that rotting tissue is removed. This will be very painful, but there is no other way to get to the fungus that has caused Mr. Deibler's condition."

Morning after morning I sat on the end of the bed dressing Russell's feet. My unpleasant task was compounded by the strangeness I experienced in the presence of this wasted, spare man. Russell would laugh when I furtively looked up at him while I tended to his feet. It continued to tickle him that my garrulous nature had been bested.

Dr. Jaffray, his daughter, Margaret, and I spent hours listening to the firsthand account of Russell's trek to the Wissel Lakes. After the first session, I came to wonder not at how much weight he had lost, but that he had survived at all.

During the first three days of the trip, the carriers forged their way upstream, and Russell claimed that in all of his travels up and down Borneo's treacherous rivers, he had never felt as uneasy as he did with these inept oarsmen, who nearly capsized them in midstream. Russell had felt troubled when first introduced to the carriers; now he knew that he should have trusted his intuition. They, like many of the New Guinea coastals, moved with lethargic, expressionless motions. Russell saw the effects of jungle life on the long-time coastal inhabitants, debilitated by dengue fever and repeated bouts of malaria where no quinine was available.

When they arrived at Orawaja, expecting a crude base camp, they found nothing more than a large, mat-walled bamboo structure with a grass roof that barely provided a place out of the rain. His carriers were reduced to seven by sickness. The three who were ill returned to Oeta with the canoe. After a hastily prepared meal, Russell and the remaining men divided the supplies among them, then adjusted their packs for the trail, anticipating an early start the following morning. Stretching out on the floor, the carriers were soon asleep. In the stillness that followed, Russell took from his pack his Bible and diary. Lighting a candle, he read, then recorded the events of the past three exhausting days.

The trail was as hazardous as any of the explorers had described. All day the lead carriers hacked away at the jungle undergrowth that obscured the crude trail. Each succeeding day carried them farther into the central highlands, with each mountain range higher than the last. They inched along narrow ledges jutting out over huge gorges cut by the Oeta River rushing menacingly below, but serving as their guide to its source, the Wissel Lakes.

The jungle-clad mountains begrudgingly gave way to ranges covered with broken-bottle limestone outcroppings, often deceptively hidden under moss. The serrated shards cut through the leather soles of Russell's field police boots.

On New Year's Day Russell awakened feeling completely depleted physically, and concerned. His carriers were few, too few. Would the supplies last? Needing encouragement for the walk ahead into this unknown wilderness, he opened his Bible and came to a portion that seemed illumined on the page: "Be strong and of a good courage; be not afraid, neither be dismayed: for the Lord thy God is with thee whithersoever thou goest" (Joshua 1:9). In His own inimitable way, with this exhortation, God fortified Russell's soul for the desperate days ahead.

Great care was exercised by those in the lead not to dislodge the stones and boulders precariously balanced beside the trail. A boulder set in motion could easily crush others on the switchbacks below. Climbing the almost perpendicular walls of the mountains, the men had to test the handholds lest a stone be pried loose and hurtle downward, injuring those beneath.

The monsoon rains deluged them. Late afternoons they made camp. Already emaciated, Russell would drop his pack, then go back to help the carriers into camp. They made the minimum of repairs on the existing bivouacs, but the shelters still leaked. These bivouacs were low lean-tos constructed of saplings pounded into the ground and laced together with rattan between the four corner posts. Sapling doors were hung on hinges made of rattan. The roof was of grass or bark. The men's clothes were always wet. They huddled around the fire for warmth, apathetically eating their rice, dried peas, and salt fish.

Picking up his diary, Russell began to read to us: "This has been a grueling day; I forced myself to eat and boil water for drinking, but I have little appetite. All day I remained at the end of the line, endeavoring to keep the carriers in sight. From their furtive glances in my direction and whispered conversations during the rest stops, I feel convinced they plan to desert. After we made camp tonight, I prayed and reasoned with them that our only hope is to remain together, trusting God for strength to go on. . . ."

That night was spent in prayer, pleading that God would keep the carriers from running away. In an early hour of the morning, Russell dropped into a troubled sleep. Noise of the carriers moving about the camp awakened him, and miracle of miracles, all were still with him.

He could sympathize wholeheartedly with the men. Six carriers had died on a former expedition over this trail. Every night the carriers, staggering under their loads, were helped into camp. Russell felt physically buffeted by the ever-increasing demand upon his fast-dwindling strength. This was not a large expedition with many police, sufficient carriers, and ample supplies; it was one man alone, exhausted, in an unfriendly jungle with seven equally exhausted carriers. However, abandonment on this ill-defined trail, one-third of the way in and two-thirds of the way to his destination, meant certain death. Even though he was able to persuade his party to accompany him further by agreeing to leave a portion of his supplies, Russell knew it was possible he could awake some morning to discover that he had been forsaken during the night and that his supplies were gone. His only recourse was prayer. To the burden of the days fraught with hazard was added the agony of pleading by night for divine intervention and protection. He encouraged himself in the New Year's promise—"Be strong and of a good courage . . . the Lord thy God is with thee"—and went on.

Not a week later, Russell realized that it was imperative they retrieve the supplies left in the bivouac downtrail, so he called a halt. He and some of the stronger men returned to bring the rest of their supplies forward by relay. After the change of pace and rest, the carriers were in a better frame of mind.

As the men climbed higher, ever higher, into the mountains, the days were warm but the nights brought the added misery of a drastic drop in temperature. Russell and the men huddled together in their blankets around the fire, trying to take warmth from one another. They were coastal men, unaccustomed to the cold of the high altitudes. They suffered, some weeping in the bitterness of believing that death hovered about them. There was no respite from the penetrating chill winds that shook their crude shelters and made a mockery of their fires. They came to dread seeing the sun sink behind the craggy ranges.

Then came that last dreadful day, the eighteenth day, a day of near-total disaster that, by the goodness and mercy of God, ended at midnight in victory!

They topped the fourteenth mountain range and emerged in Kapauku country. The terrain was comparatively flat, and a trail wound between sweet potato gardens. Sometimes the men toiled through mud up to their hips.

About three in the afternoon, the trail led to the river's edge, where they found canoes that were evidently used by government parties. They had not paddled far before they realized it would be suicidal to go on. A storm upriver on Lake Paniai caused high waves and dangerous currents where the lake, in its feverish thrust toward the coast and the sea, funneled out into a narrow, high-walled channel to become the Oeta River. Six miserable hours they

waited in the canoes for the wind and the waves to die down. By
nine o'clock, all of them were shivering from the cold, and Russell
felt he could wait no longer. Paddling against the current in the
river warmed their blood. With grateful hearts, they cheered when
they swept out of the river channel into the larger body of water.
They still had to cross the corner of the lake to reach Enarotali,
the government post.

Faint lights could be seen on the distant shore. Just when they
began to relax, the ordeal seemingly over, Russell's canoe struck a
submerged rock and overturned. Every person and piece of gear
plummeted into the tumultuous, icy lake. The piercing cries for
help from the carriers awoke the government personnel. Russell
and the carriers swam madly in the current, and all managed to
scramble aboard the second canoe. They bailed frantically, for the
canoe, now severely overloaded, was shipping water. Lanterns appeared and kind hands helped to pull the canoe onto the lake-
shore. Though water-soaked, every man and piece of luggage was
eventually salvaged.

At midnight on January 13, 1939, Russell, a lone pioneer missionary, stepped foot on the land, like Joshua of old, to lay claim
to the territory and all of the primitive tribes scattered throughout
the interior valleys and ridges, anticipating the time when all would
hear the Gospel of Jesus Christ.

At dawn Russell peeked out of the makeshift government hut
only a little rested but eager to meet the native people. Dozens of
dark brown, pygmoid people with full lips and broad noses swarmed
around the post during the morning, curious to see the newcomers. One brave man quickly offered his bent index finger to Russell, who stood back and watched as the patrol officer reached out
and clasped the man's finger between his own bent index and
middle fingers. Both then pulled their hands apart and downward,
producing a sharp snap. This was repeated many times—the more
snaps and the louder, the more certain the bond of friendship.

Feeling brave with his newly learned greeting, Russell mingled
among the people. Their wiry hair was caked with dirt and ashes,
and their small bodies were smeared with a combination of mud
and pig fat that closed the pores and substituted for clothing. The
men's complete attire was a phallocrypt gourd, which hung from
a string around the waist and was held in place by another string

around the scrotum. Little girls wore grass skirts; women wore brief string or rope skirts. The women and girls carried net bags that hung from their heads, functioning as a carry-all by day and a Stone Age version of thermal underwear by night, as they stretched them around their small bodies.

A hole was pierced in the nasal septum of both boys and girls at an early age. A straw was inserted in the hole until the perforation healed; then a small reed replaced the straw. Periodically the reeds were exchanged for one slightly bigger, until the hole was large enough for the men to wear a pig tusk or a piece of bamboo and the women a reed or stick pointed on either end. All had pierced ears, not for earrings but for an occasional quill from a cassowary feather; or more practically, the holes functioned as a storage place for roughly crafted cigarettes or bamboo pipes. Without trousers, shirts, or pockets, they found ingenious ways to store necessary items!

Necklaces fashioned from shells or the teeth of dogs or rats were worn by both sexes. Wigs fabricated from cassowary feathers commonly covered the balding heads of the proud but aging men.

The Kapauku houses were built of hand-split planks, pointed on one end and pounded into the ground for a crude framework. The trees were felled with stone axes; the trimming or shaping of the planks was done with stone adzes. The walls were a single thickness, and the roof was of tree bark. The floor was bare earth, with three stones forming the fireplace. The householders slept around the fire in a tight cluster with the pigs, especially the piglets that were being suckled by the women when a mother pig had died.

More than thirty varieties of sweet potatoes were grown and tended by the women in the surrounding lake area. In the absence of cooking utensils, the sweet potatoes and edible greens were sometimes steamed with hot rocks in an outdoor pit, but they were more often roasted in the hot ash under the coals in the indoor firepit.

Pigs, killed for spirit appeasement, rats, crayfish, tadpoles, birds, caterpillars, wichety grubs, bee and wasp larvae, grasshoppers, stinkbugs, and other insects supplied interesting varieties of protein in their diet.

The Kapaukus exchanged cowrie shells as money. Russell found that a large, fat pig or a young, strong wife could be bought for the same price—a string of from forty to sixty *yo* (the old, thin cowries). It dealt my female pride a blow to know I would be considered no more valuable than a dirty, fat hog!

Dividing his time between firsthand experience with the Kapaukus and the friendly, knowledgeable government personnel, Russell was able to ascertain what materials and arrangements were available locally and what would have to be brought in to establish a permanent mission station.

Russell was at the point of complete physical exhaustion. Understandably, he faced the arduous trek back to the coast with reluctance. During his stay the patrol officer became very ill with fever, and when the governor dispatched a navy plane to evacuate him, the officer requested, against standard procedure, that Russell be allowed to accompany him. "Look at his feet! He'll never make it back to Oeta alive!" Three hours later they deplaned in Manokwari. From there he booked passage on an interisland steamer to Ambon, where Mr. Post joined him, and then to Macassar. That providentially arranged plane ride saved Russell the impossible trek to Oeta and exactly one month of travel time.

One morning after Russell had told the final episode of his story, Dr. Jaffray walked into the bedroom and saw me tearing the dead tissue off Russell's feet, the blood and pus running. A wave of nausea passed over his face; then he turned and without a word abruptly left the room. He closeted himself in his bedroom, and when I called him at noon, he said he would not be out for lunch. About four that afternoon, he walked out and laid a manuscript on the table in front of me.

I picked it up and read the editorial for our field magazine, *The Pioneer.*

> This morning I looked at the bleeding feet of a missionary, saw his wife tending them, saw the blood and pus running from them and thought to myself, "What a nauseating sight that is!" But, as I walked from the room, the Lord kept saying to me, "Oh, but to Me they are beautiful feet!"
>
> Then I remembered—"How beautiful upon the mountains are the feet of him that bringeth good tidings"—good tidings to men and women like those in New Guinea who sit in darkness and in the shadow of death. Someday it will all be over. Someday the tired, bleeding feet

of the missionaries will for the last time cross those broken-bottle limestone mountains. Someday for the last time they will go down into one of those newly discovered valleys. Someday for the last time they will speak the message of redemption through Jesus Christ our Lord. Someday that last one will turn to Jesus. Then the clouds will part asunder and our Savior will be there.

Reverently, I laid the manuscript on the table and lifted my tear-filled eyes toward the east. I knew that soon I would join the long line of intrepid missionary pioneers who had walked into the unknown to lift up His ensign on the mountains and lay a claim for the Lord.

From that day on, although hurting at the pain Russell was experiencing when I dressed his feet, I thrilled to think my Lord was looking at them, saying, "To Me they are beautiful feet!" It served to increase my desire to join him when he returned to Wissel Lakes. I was impatient. Dr. Jaffray had shared his vision for the day when it would be over. I just wanted it to *start!* When Lord, when?

CHAPTER TWO

Our front bedroom became the "oval office" where every facet of setting up a station in New Guinea was the main topic of discussion—the problem of carriers, transportation, supplies, and housing. Russell was not yet mobile.

One morning Dr. Jaffray walked in to announce, "Russell, I have the answer! While in prayer, the Lord reminded me that the Babo Oil Company and all the other expeditions to New Guinea used Dyaks. We need Dyaks!"

Immediately, we all knew he was right. They were without peer when it came to river or jungle work, and where was a better place to get them than from the area where Russell had worked in Borneo?

Dr. Jaffray's letter to the resident missionary brought a speedy reply. In record time twenty sturdy Christian Dyaks, their muscles rippling when they walked, arrived in Macassar by interisland steamer. Mr. Post was dispatched to Ambon to make arrangements for housing the Dyaks, and three teachers from our Bible School upon their arrival.

In the days following the doctor's okay for Russell to be on his feet, I saw him only at meals and bedtime. A contract was drawn up and notarized for the Dyaks. They and the teachers had to be outfitted. Chinese artisans crafted lightweight trail tins with water-proof lids, tailored khaki shirts and trousers to fit Russell's thinner frame, and crafted his boots with cat's claws fitted to the soles. He purchased and packed food for twenty-five with the help of the Dyaks, who were staying in the Bible School dormitory. By March 5, all was in readiness.

Padjak, from the same area as the carriers, was invaluable as their translator. He, Saragih, a well-educated school teacher from northern Sumatra, and Pattipelohij, a school teacher from Ambon, were teamed to help in the ministry. They were very different in every way, yet they complemented one another and made a terrific team.

The farewell service held at the Tabernacle was an hour long remembered. The Dyaks stood on the platform, attired in breech-clouts and pill-box hats of finely woven fiber, and sang songs of praise to God in their own several-part harmony and language. A thrill went through the audience. I thought I'd never heard such beautiful music. The Dyaks' faces manifested gratitude when the whole congregation solemnly raised their hands in a pledge to pray for them as they went to New Guinea.

Saying goodbye was much easier this time, knowing that Russell would have trustworthy, capable carriers and companions.

At Ambon they transshipped and were joined by Mr. Post, then continued on to Oeta. The Dyaks adeptly handled the coastal canoes through the first rapids, cutting the travel time to the base camp in half. There they sorted and packed supplies, then fashioned slings in which to carry the trail tins. The machete, or long knife, was as much a part of the Dyak's attire as his loincloth. They were hand-forged and very sharp. Wherever the Dyaks walked, swinging their machetes, the jungle receded from around the bivouacs and along the trail.

The supplies were transported over the trail by relay. This meant that Russell and the carriers traversed the trail twice to get everything in from the base camp.

The Dyak carriers sought out easier and more direct routes, felling trees for bridges or fashioning ladders for scaling the mountain walls, thus saving days on the trail. The Dyaks also hewed and burned out three splendid canoes shaped for speed and safety on Lake Paniai and the reticulated rivers of Kapauku country.

Once at Enarotali, housing was the prime concern. Four Dyaks were kept busy preparing materials for building an initial structure, sufficiently large to serve as storehouse and living quarters for the Dyaks and mission personnel. Trees were felled for the framework, tree bark was flattened and stacked for roof covering, bamboo mats were woven for walls, and rattan was collected for tying the building together. Everything was locally available, with the exception of a few nails and the sheets of isinglass. The total cost of materials was approximately twenty dollars U.S. 1939.

The Dyaks, accustomed to rice and fish in plentiful supply in Borneo, found the sweet potato a poor substitute; some became sick. It became apparent that someone must make a trip back to Ambon for further supplies. Russell was elected.

He walked in driving rain for five days, from 6:00 A.M. until 5:00 P.M. The trail was dreadful but the day he left the base camp in a canoe, in a tropical downpour, was even worse.

Arriving in Oeta, he was informed that the steamer was not scheduled to stop that month. "But," he said, "I prayed, and two days later I was aboard ship en route to Ambon."

Since Russell would be in Ambon about a month before the next steamer to Oeta, I was granted permission to join him. In the three months since I had seen him last he had gained weight and now weighed about 140 pounds. I stuck with him every glorious moment, and I'm sure there wasn't a full twenty-four-hour day all month or it wouldn't have ended so quickly. But end it did, and I found myself once more watching a steamer carry him away—for how long this time, only the Lord knew.

I came down with dengue (or "breakbone") fever, but knowing that the rash would not appear until the fifth day, I caught the next steamer for Macassar. When the cabin boy opened my door, he saw what appeared to be a bad case of smallpox. With a gasp, he slammed the door shut and disappeared. Some hours later, the first officer came to see me. Recognizing the symptoms of dengue fever, he said I should remain in the cabin until the rash disappeared. He sent my meals to me, and I was quite recovered before docking in Macassar.

Dr. Jaffray had been very ill during my absence. He had been in a coma for several hours. The local doctor diagnosed a "peculiar type of malaria." This seemed to be the cover-all for any illness the doctors weren't sure of. Dr. Jaffray was much better, but he was remaining at Benteng Tinggi until stronger.

Margaret Jaffray and I joined the other missionaries in Benteng Tinggi for the remainder of the school holidays. Very late on the night of September 3, 1939, we were startled awake by insistent knocking on the door. It was a Dutch patrol officer and his aide with the news that an hour earlier, England and France had declared war on Germany.

Holland mobilized her troops in the motherland; however, she hoped to remain neutral as in World War I. If she were drawn into the European conflict, what of the Indies? Would the surrounding British and French colonies and possessions bring the war to the Far East? Precautions were being taken, and the Marine planes serving the expedition in New Guinea were withdrawn to Java.

Government officials in Enarotali were ordered to be prepared to abandon the post and proceed to the coast at a moment's notice. The American consul ordered all Americans to have their passports up to date. From that time on, the conflict in Europe was never far from our thoughts, and constantly in our prayers.

In late September I had returned, as had the other missionaries, to our ministries in Macassar, when a cable came from the mission's consul in Java assuring Dr. Jaffray that permission to use the mission's plane in New Guinea would be granted immediately.

Dr. Jaffray broke the wonderful news to me as soon as he heard: "Do you need some help packing, Lassie?"

"Thank you, Dr. Jaffray, but this being the eighth time I've packed—and unpacked—in the last few months, I'm quite adept at it."

I packed with alacrity and care. Cardboard boxes would be used for transport by air. In the sultry weather, I had to check the boxes daily for mildew and vermin. Cardboard was no protection against the ever-present rats and cockroaches.

The promised permit, to use our plane in New Guinea, never came, so a few weeks later, I returned the few things I had chosen to take with me to New Guinea to the five metal trunks of wedding gifts. Handling the beautiful gifts was always a time for remembering our dear families, and the multitude of wonderful friends God had given Russell and me. As long as the plane permit was held up and the trail was the only means of reaching the interior, our wedding gifts would remain in the trunks with a liberal supply of roach and rat killer.

In October, while Mr. Post was in Ambon, Russell and three Dyak carriers explored the newly discovered Kemandora Valley to the east of the Wissel Lakes, which had an estimated population of twenty thousand. After three days the police patrol that Russell accompanied aborted its plans; but rather than suffer defeat, Russell continued with his three carriers, and for several days they traveled through mud up to their knees, often falling waist deep into it. Many days a streambed served as their trail. The Moni people welcomed Russell and the Dyaks and brought fruit and vegetables to them.

The day before they headed back toward the Wissel Lakes, Russell traveled to the eastern limits of Moni land with Moni boys as guides, allowing the Dyaks to rest. Standing atop a very high mountain, he gazed into a labyrinth of valleys, mountains, and passes and wondered how many other tribes were locked in beyond where his eyes could see. He felt a bit like Abraham overlooking Canaan, who heard God say, "Lift up now thine eyes, and look from the place where thou art northward, and southward, and eastward, and westward. For all the land which thou seest, to thee will I give it" (Genesis 13:14–15). Russell heard the Lord speak to him on that high place: "Other sheep I have, ... them also I must bring." Yes, there were many more undiscovered tribes, and much land to be possessed.

While Russell continued to explore the uncharted areas far from the news of the outside world, a German freighter, flying a swastika, lay at anchor in Macassar harbor all of October and part of November. Were the Germans awaiting orders or just enjoying the pleasures and protection of a neutral port? we silently wondered.

Again Russell celebrated Christmas on the trail, while I was hundreds of miles away at Benteng Tinggi with the Macassar missionaries. Russell and I had been apart for seven more long months. The Jaffrays, Grace Dittmar (a newly arrived missionary), and I went caroling at four in the morning and awakened everyone on the hillside with our loud rendition of "Joy to the World." Even as we sang, the wonder of God's great love seeped into all the crevices and voids of my inmost being, giving me strength to face the dark, looming clouds of war and further separation from Russell.

A Christmas letter arrived just days after the holidays. Russell had accompanied the Dyaks to the coast to see them safely aboard the government steamer returning to Borneo. Their six months were over, and as they left, Russell wrote, "I shall miss them. We thank God for the Dyaks. We have worked with them side by side for so many months that I shall be lonely without them."

The letter recounted his visit with the chieftain at Enarotali. Idantori squatted in his hut and listened attentively as Russell tried to explain the love of God, and how Jesus had died for them. The chieftain was spiritually hungry, he admitted, but if Jesus had died for Russell, then he hadn't died for them. They were people and Russell was a spirit—one who had come from the spirit world beyond the mountains.

"I am not a spirit!" Russell protested. Idantori, to prove his theory, took his Stone Age calendar off of the wall and began to count the rattan knots—eighteen in all. The first knot had been tied when the patrol officer and the police arrived. Each knot thereafter represented the passing of each "moon." "You arrived, and later others followed. You are all men," Idantori insisted. "None of you has a wife or children. If you are not spirits, who gave birth to you?"

Russell countered, "But I told you before that I have a wife."

"Where is she?"

"Macassar," which to Idantori probably sounded like a good name for a spirit place.

"Why has she not come here with you?"

With a limited vocabulary, Russell found it difficult to explain permits and government regulations, so he lamely added, "Our chieftain says, 'no!' "

"If your chieftain is so bad in his stomach that he won't let your wives come, get rid of him!" Idantori reasoned.

"Many men have died on the trail," Russell protested, to which Idantori responded, "Then your wife would have made it. Wherever we go, the women follow and carry the loads."

Idantori eventually concluded that if Russell were speaking the truth and there was a wife so weak that she couldn't carry his supplies, then he would send Kapauku men downtrail to help her!

Meanwhile, others of the local tribe decided to learn firsthand if the government personnel were truly spirit people. They reasoned that if their arrows killed these intruders from beyond the mountains, they must be human. However, if the arrows passed through their bodies and they did not die, then they were spirits. A small government party that had gone across the lake was ambushed by several hundred natives. Some Kapaukus were killed when the police opened fire in self-defense.

"Lord," I prayed, "if those people are ever to believe and understand about You, women will have to go there." The moment I spoke those words, an assurance filled me. God spoke clearly in the silence. I threw the letter into the air and yelled, "I'm going to New Guinea!" Scooping up the pages, I dashed out to find Dr. Jaffray.

"Dr. Jaffray, read this! I'm going to New Guinea!"

"Lassie, I've known that for several days. I also had a letter from Russell giving his consent for you to go by trail. I've been waiting for the Lord to show you."

Dr. Jaffray and Margaret took me to the ship on January 23, 1940. Our belongings, packed as if by a seasoned professional, were stored below deck in the cargo hull. Dr. Jaffray commended me into God's care; then he said, "Remember, Lassie, for centuries the enemy has held these people in darkness. You will now experience satanic opposition such as you have never known. Until Russell's first trip, no one had ever invaded his territory to challenge him, but don't be afraid, for he is a defeated foe, undone by Calvary. Never forget that greater is He that is in you than he that is in the world."

Pondering these words, I stood at the rail and waved until the ship moved out beyond the breakwater and the Jaffrays were out of sight. They had become like a father and sister to me.

I joined Viola Post in Ambon, where she had remained since her October visit with her husband. At the crack of dawn on February 5, we were up and moving about. Supplies, carefully packed in tins or crated for repacking in Oeta, were put on board; then we embarked on the small government steamer.

Almost every day we stopped at a different island to discharge cargo for remote government posts. Visiting these islands, tiny dots anchored by their coral moorings to the ocean floor, was like being transported into yesteryears of history. There was the ancient fort of Banda, built by the Dutch when they expelled the Portuguese in the early 1600s. Its ancient cannon, like a Cyclopean eye in the crenelated structure, watched the harbor. The old, old churches evidenced the labors of early Dutch missionaries. A white linen cloth and silver communion service, over 100 years old, were lovingly cared for and still being used. The clove, nutmeg, mace, and coconut plantations were a delightful experience for my olfactory nerves. The marine gardens beneath the sea of this area were virtually unrivaled. I was most intrigued by the pearl fisheries. The Japanese divers were able to descend to amazing depths without the use of diving equipment. They were soon to be recalled to their homeland with in-depth information about the Banda and Ceram Seas and the surrounding islands.

We were traveling at the worst time of the year in terms of the weather. The government boat was small. While rolling from side to side, it pitched into, onto, and off of the towering waves. Thank God I wasn't prone to seasickness. As we turned south and east along the neck of the island, the weather steadily deteriorated. We knew that unless there was a change for the better, we would not be disembarking at Oeta the following day.

Because of a large sandbar guarding the mouth of the Oeta River, the government steamer had to drop anchor at considerable distance from shore. Lowering the cargo and passengers into lifeboats during a storm of this proportion was a risk the captain did not care to take. The alternative was to continue on to the next scheduled stop: Merauke, a government post near the border between Netherlands New Guinea and Papua. We were assured that there were no objections to our remaining on board. During the return trip it would surely be calm and they would be able to land us.

The captain and officers were more than gracious, but there were so many factors involved that Viola and I were emboldened to ask the Lord's intervention. Our husbands would already have arrived on the coast. Because there was no radio in Oeta, it was impossible to inform them of any change in schedule. Two of the evangelists and a contingent of Kapaukus would be midway downtrail, to wait at Orawaja. There was no way to turn them back. Housing and feeding so many on the coast would be expensive and a near impossibility. The danger of the Kapaukus contracting malaria also had to be considered. Never before had they ventured to the coast. The urging to find out if we were spirit people or humans, like them, had moved them to volunteer for the treacherous trip.

God was our only source of help. We prayed earnestly that He would still the wind and calm the troubled sea. By morning the sky had begun to clear, the wind to abate, and when we came abreast of Oeta, the ship dropped anchor in the glassiest sea I have ever seen in all my trips around the world. Today's seas are still subject to the Christ of Galilee. In about an hour and a half, we and our cargo were lowered into a lifeboat and transported around the sandbar. Excitement mounted as I glimpsed Russell standing on the jetty; once more we were to be together, this time in New Guinea.

Three days later, in dugout canoes with coastal oarsmen, the Posts, Russell, and I started out for the base camp.

When we arrived at the base camp, the supplies were quickly unloaded and stored in the large shed, and the carriers left immediately to return to Oeta. We were sorting supplies when we heard loud yodeling—"hoo-hoo"—echoing through the jungle.

"The Kapaukus!" Russell yelled.

Padjak and Pattipelohij arrived with the Kapaukus, sent along by Idantori. They ran into the clearing in a circular formation, "hoo-hooing" in a dissonant two-part harmony.

Stopping after a few minutes, they called, "Where is your woman?" Russell pointed to where I had run up on a pile of fallen tree trunks to see them better. They surrounded me like a flock of chattering magpies, examining the whiteness of my skin, shoving up my sleeve to see if the color was the same all the way up. Some gave my arms the pinch test, nodding to one another as though to say, "Yes, see, it has the feel of real flesh."

It was a moment pregnant with deep emotion. These dark-skinned, pygmoid people were New Guinea's people—my people.

The small kerosene lanterns had been lighted, so I sat down watching the Kapaukus divide their sweet potatoes into equal piles and place them in their tins. To the approximately forty pounds in the tins, Russell and Walter added from our supplies enough to make up a sixty-pound starting load.

I was intrigued, watching them tie their tins with rattan pulled from the nearby trees. The handles they fashioned were tried, then shortened or lengthened until the tins hung from their heads in the most comfortable position for carrying.

I asked Russell if sixty pounds were not too much. "No," he explained, "it's a diminishing load as they eat their potatoes. Traveling with them in the Wissel Lakes area, I learned that they are an amazingly strong people; and don't forget, they're accustomed to the mountains, the altitude, and the cold."

What a wonderful answer to our carrier problem. With the Dyaks gone and the coastals physically unfit for the rigors of the trail and the cold, we would never have been allowed to make this trip without the help of the Kapaukus who had volunteered to help us "weak women."

"Russell, what are they saying? They keep looking in our direction."

"Oh, they're probably wondering if they'll have to carry that weak, skinny woman before the trip's over, and—"

"Russell!" I bleated, "You know how strong I am!" Then I saw his grin. "What are they really saying?"

"They're wondering why you aren't wearing glasses like the rest of us, and they've never seen blue eyes before. They're as curious about you as you are about them."

As soon as their trail tins were ready for the trail, the Kapaukus curled up on the floor to sleep. I crawled in under the mosquito net that Russell had hung for me; I thought that I was much too excited to sleep. But I did.

The first day on the trail, the torrential downpour came. We climbed in the driving rain, determined to reach one of the bivouacs, but didn't make it. So we stretched a tarpaulin Viola and I had waterproofed, but it proved no protection against the deluge. It was a miserable night for all of us, and before dawn we were thankful to be on our way.

All that long day, thoroughly drenched, not daring to stop lest we become chilled, we kept moving until we reached one of the bivouacs, where we camped for the night. Because it was still raining when we reached the campsite, there was no dry grass to cut, so I snuggled down into the luxury of the much-used dry grass on the earth floor, thinking how wonderful it was to be so comfortable. About midnight the rain stopped.

We found the single-log bridges that spanned the gorges to be great time-savers. Each had a piece of rattan for a handrail. Anything that lies dormant in the jungle for even a short time becomes moss-covered and slippery, so I was grateful for the cat's claws on my boots. I reminded the Lord frequently of his promise: "None of his steps shall slide." To fall from these logs would have meant broken bones—possibly death.

If the ladders improvised by the Dyaks, during their time in New Guinea, had rotted, new ones were speedily fashioned, as raw materials were in ample supply. We threw them up against the perpendicular walls of the mountains, saving hours and energy.

We changed to dry clothes for sleeping, but the following morning we dressed in our washed, but damp, gelid trail clothes of the previous day. This, to me, was the most unpleasant part of the trail.

If the Kapaukus thought I was lagging behind, they would turn to each other in mock seriousness and say, with a toss of the head in my direction, "Look at her, the old thing—no legs on her at all!" My youthful pride couldn't ignore their teasing. Of *course* I could walk faster—which was exactly what they wanted.

All the morning of the final day, the carriers, Padjak, and Patti-pelohij kept urging me on: "Today we will be in Edupa, the first village in Kapauku land." Approaching the top of the final mountain ridge, I could hear the carriers, who had gone on ahead, yelling, "The women are here! The women are here! They are *not* spirit people."

Cresting the summit, I looked down into the valley and saw men, women, and children running out of their gardens or hurrying out of their huts. All were heading toward the mountainside. Half of them yodeled "Hoo!" and then the answering "Hoo!" echoed back an octave lower from the rest of the crowd.

Their excitement was infectious. Those long months of sitting on the coast in Macassar, the separation from Russell, the packing and unpacking, the exhausting trek over the rugged terrain—all these were behind me now.

I raised my hands, waving to the people. My cheeks streaked with tears, I started running down the mountainside, singing at the top of my lungs, "I'm home! I'm home!"

The first Kapauku women met me halfway up the mountainside, each bearing a gift—a roasted sweet potato. I couldn't hold them all, so I sat down on the grassy slope as they continued to pour their presents on my lap.

Then the chieftain of Edupa, Kepala Edupa, came up to me, demanding, "Are you a woman or not?"

"Yes, I am," I replied, as he leaned close to scrutinize me carefully, comparing my hat, my clothes, and my boots to Russell's, who had just come over the top of the mountain.

Kepala Edupa turned down his lower lip and, with evident mistrust, spit out, "No, you are not!"

"Yes, I am a woman," I declared, and took off my hat and removed the pins, letting my hair tumble over my shoulders.

Kepala Edupa dropped his bow and arrows, bit his index finger, clicked his gourd, and exclaimed, "Yes, you *are* a woman! Truly, no man ever had hair like that!" He took hold of a lock, giving it a yank to make sure it was attached to my scalp.

The chieftain felt through the pile of sweet potatoes, searching for one that was still warm from the coals of the fire. Finding one, he took the sweet potato in his hands, which had never been washed—not on purpose, anyway—rolled it, blew on it to remove the dust and ashes, then peeled it with the paring knife that had grown on the end of his thumb. Rolling it between his hands again, giving it a second blow test, he handed it to me, saying, "Woman, eat the potato."

I took the sweet potato from him and, though I had been studying word lists from Russell and thus knew greetings and understood what Kepala Edupa had just said, I didn't know sufficient to say what was burning inside of me. I said aloud in English, "Someday, by the grace of Almighty God, I will sit down and eat with you anew in my Father's kingdom." I love sweet potatoes! I ate it with relish.

I sat encircled by the children and the women until Russell joined me. He was no stranger to them, so after a brief visit we went on with the carriers, with whom I shared my sweet potatoes. Reaching the canoes left on the river's bank, we waited for Viola and Walter, who soon joined us.

An official welcome awaited us at Enarotali. Dr. J. Victor de Bruyn, the Dutch patrol officer, invited us to his quarters for a cup of tea and fried sweet potatoes. It was apparent that he was held in high esteem by the local people.

Then Russell took me home, our very first home. It was beautiful to me. There were two rooms: a livingroom and study-bedroom with woven bamboo-mat walls and floor. The bed was made of pit-sawn planks, as was the counter across one side of the bedroom.

What a wonderful idea, using isinglass for windows to keep out the cold and let in the view. Russell had chosen a lovely spot for the house, on the hillside looking across Lake Paniai to the tree-clad mountains behind which the sun was just setting. The magnificence of the sunset mirrored on the lake was breathtaking.

The whole lovely home was marvelously air-conditioned—the air came through most anywhere. Upkeep was simple: if the roof leaked, we went to the jungle, got another piece of tree bark, and slipped it in. Often other things could slip in, too, like the three-foot-long forest dragon that was hanging on the wall above my head when I awakened one morning.

The kitchen was a lean-to with an open-wood fire. I made my own oven from a five-gallon kerosene tin cut in half lengthwise. With a piece cut from another tin, I devised a rack on which to set my bread pans. When I set my oven on a firepot made of local clay, I had a functional, custom-made, top-of-the-line, open-fire oven.

Starting for the kitchen, I almost stepped on the newest member of our family—a little Kapauku boy. He was juggling live coals and grinned at me, saying he was going to start my fire. After a kettle of water was set over the flames, Imopai informed me that his mother was dead and he was my boy. I was willing. I couldn't resist his grin.

Scarcely were our trail tins stacked in the *godown* before our official welcome to Enarotali from the local people was celebrated.

Part of our responsibilities as pioneer missionaries was making contact with other villagers living along the rivers that flowed into the Wissel Lakes. We traveled the waterways in our Dyak canoe, moving from one village to another, some of which Russell had visited before, until we reached the outermost Kapauku villages, east of the lakes.

One evening Russell connected our twelve-volt, battery-powered radio, then switched it on to catch the BBC London news. We leaned over the radio, not wanting to believe the shocking news, that the Nazis were invading Holland. Russell ran down to the government post. They, too, had heard the news only hours earlier. Our thoughts and prayers turned to our dear friends in Holland and to that tiny country pitted against a monstrous war machine, Nazi Germany. It was the tenth of May 1940, my twenty-third birthday.

In just five days Holland fell. It had been infiltrated, betrayed by the enemy within. This meant dire days for those in Holland, austerity abroad, a "tightening of the belt," a determination to endure until the motherland was free of the tyrant. In some areas it spelled retrenchment. The royal family had safely reached London, and Queen Wilhelmina spoke courage into the hearts of her loyal subjects in the colonies. With German U-boats and submarines operating in the Indian Ocean, the Java Sea, and the Macassar Straits, the long arm of the conflict in Europe was beginning to be felt in our peaceful islands.

Many times during each day, I would find myself saying, "Thank you, Lord, it's so wonderful to be here!" We were striving for mastery of the language so that we could leave believers behind to encourage and comfort one another.

The people, with the help and supervision of the teachers, constructed a building that weekdays served as school house and Sundays as a church building. Both Pattipelohij and Saragih were good teachers, instructing their pupils in speaking, reading, and writing Indonesian, plus basic arithmetic.

Sunday services were well attended, both by people and by flies. Each churchgoer arrived with his or her private swarm. One Sunday the flies were more pestiferous than usual. Degaamo, the village "bad guy," was sitting well up front, swatting flies to the right, to the left, to the rear. Finally in desperation he jumped up and, pointing a grubby finger at the preacher, Mr. Post, demanded, "Do you have flies in heaven?" The preacher had never had that theological conundrum thrown at him before; he couldn't think of any appropriate chapter and verse! Receiving only shocked silence as his reply, Degaamo said in disgust, "Well, if you do, I don't want to go there!" and stomped out.

Daily in the afternoon, with our picture roll, we visited from house to house, sharing and witnessing. I recall vividly one wizened little old lady who prayed with me regularly for others after she came to know the Lord. Her skin was wrinkled and gray with the ashes and dirt of her many years. Her hair had been rubbed off where the handle of her nets had rested during the long days of her life from young womanhood to old age. Her teeth were still firm, as were those of most Kapaukus. Her eyesight was dimming, but to me she was beautiful. Always I was welcomed with an ear-to-ear toothy smile, a sweet potato, and often crayfish, which she had roasted on the coals of the fire.

After a time of prayer, we shared the repast. She peeled and blew my sweet potato. This I ate while she made ready the crayfish. Knowing that I didn't particularly care for the bitter flavor of the yellow, liverish substance left on the tail after the head was broken off, she sucked it off before handing the tail to me to shuck and eat. With her long thumbnail she was well equipped to dig the "goodies" out of the shells, so she always kept the heads from my apportioned number of crayfish as well as her own. If I had gone alone to the village, either my elderly friend or Idantori's youngest wife and sister-in-law escorted me home.

Outposts that were remote or yielded little to advance the war effort were being closed. All too soon the word came: "Enarotali must be abandoned." We begged permission to stay, but without police protection it could not be allowed. All police were to be posted to Ambon, and the services of the government steamer would be discontinued.

The government would be held responsible for our safety, especially because women were involved. This I could appreciate, but it was so grievous to be wrenched from our home and our people. How I needed the Lord's words for courage and comfort: "I know the thoughts that I think toward you, . . . thoughts of peace, and not of evil" (Jeremiah 29:11).

We started packing, but every free moment was spent with the people. We tried to explain that our chieftain had said we had to leave. Idantori just *knew* we should have gotten rid of that chieftain with the bad stomach!

On the morning that the governor radioed Dr. de Bruyn that on a certain date the *Albatross,* the government steamer, would call at Oeta to pick up all personnel from Enarotali, we knew that immediate evacuation was imperative.

Dr. de Bruyn, no less grieved than we, drew up the plan. He and a party of police would go first to attend to the procuring of canoes from Oeta. The mission party would be followed by the remaining police and coastal prisoners in two contingents. The police commandant with the last group would carry out the government radio and generator. When we said goodbye to Dr. de Bruyn, he teasingly said to me, "If you make it to the base camp in five days, I'll kill all the goats and we'll have *saute kambing* 'goat shish kebabs'." Taking the challenge, I laughed and said, "You're on!"

Coming up from the *godown* with our trail clothes and boots, I saw Imopai, my faithful bearer of hot coals, sitting on our porch with several young friends. Someone had given Imopai an old straw hat. He was never without it. He and I were both terribly distressed at our imminent departure. He was my boy; I had become his mother. I wanted so much to know before we left that he understood why we had come to live among his people. As meticulously as I could, I explained about God's love.

"Oh, Imopai, do you understand?"

He had been staring at his hands. Suddenly he looked up, understanding showing in his big brown eyes. "Yes, Mama, I have listened. *Jetoti,* 'Jesus,' died for me." We bowed our heads while Imopai prayed and God heard.

When we left the next day, Imopai accompanied us over the first mountain range. Finally I sadly said, "Imopai, you must turn around and go back now." I held his hand in a tight clasp; I could say no more. Tears were too near the surface. He stopped, and I walked on down into the valley. When I turned, there he was, standing on the mountainside. I saw the old straw hat and a little boy, so alone, silhouetted against the afternoon sky. I could tell he was crying. Feeling that he was "too old to cry," he wiped his face with an angry gesture, then brushed the tears off on his hip.

Finally he called, *Mama, egaa kedaa!* "Return quickly!"

The wind reaching up from the valley carried my answer: *Imopai, kojaa tou.* "Remain in peace. As soon as we can, we will return."

At the bend of the trail, I looked back. Imopai was gone. Sitting down, I wept for my boy, my son in the faith. "Dear God, please take care of him, until I can come back." I little realized that it would be nine years and a war away before I'd see him again.

My trail boots, being constantly wet, came apart at the seams, so the last two days I had them tied to my feet with rattan. Nevertheless, I made it to Orawaja in four days. True to his word, Dr. de Bruyn killed the goats and we all enjoyed goat shish kebabs.

Waiting in Oeta for the rest of the government party to arrive, we went on a spending spree—buying lukewarm Orange Crush.

On the appointed day, the *Albatross* arrived, transporting us to Ambon. German U-boats and submarines had been reported in East Indies waters. A strict blackout was observed as we shipped from Ambon to Macassar. Frequent lifeboat drills were constant reminders that the conflict in Europe had now engulfed our peaceful islands.

CHAPTER THREE

We'd been back on Celebes working in the Bible School for nearly five months when on the eve of the annual conference we heard the exciting news: the post at Enarotali was to be reopened and the missionaries had permission to return. In the eventuality of the war in the Far East spreading south, an interior post in New Guinea would be strategic. Already Japan had military tentacles around the throats of China and Indochina. Russell and I allowed our hearts to soar in anticipation of heading back to New Guinea as soon as the conference ended.

With the mission's expansion into New Guinea and Malaysia, the administrative load continued to grow, and Dr. Jaffray felt the need of a younger man to be his assistant. Despite his physical condition (he was diabetic), Dr. Jaffray rose at four o'clock to write for the Chinese *Bible Magazine,* of which he was editor.

We were advised to be much in prayer that God would reveal His choice for the assistant field chairman. No recommendations were brought in by the allocations committee, and no nominations were made from the floor.

Slips of paper were circulated, and we wrote down the name of the person the Lord had laid on our hearts. These slips were collected and counted by the tellers. Presently they returned and handed a paper to Dr. A. C. Snead, who had come from the New York headquarters, at Dr. Jaffray's request, to chair the conference. We all waited with bated breath. I was sure I knew who it would be; he would like it and would be a good choice.

Instead, Dr. Snead read Russell's name, and it was like a death knell. We were asked to rise while they sang, "Bless them, Lord, and make them a blessing." We both burst into tears; I had to get out of there. Something was very wrong. We had pioneered the New Guinea field; it was now to be reopened. "Lord, this just can't be! They've made a mistake in counting the ballots!"

Russell was explaining things to the Lord also, when Dr. Snead and Dr. Jaffray came out to find us. "I think you should know that every ballot had your name on it, except yours and Darlene's. I believe you can accept it as from the Lord," counseled Dr. Snead.

"Yes," said Dr. Jaffray. "You were also my choice, and it's significant that you have the unanimous backing of all your fellow missionaries."

Another missionary came up to say, "When I saw you crying because you didn't want the office, you wanted to return to New Guinea, I said, 'That's the measure of the man, the kind we need for assistant chairman.'"

The Lord reminded me of my commitment to go "anywhere, Lord." My longing to return to New Guinea, to Imopai and others, I surrendered to the Lord, put it in my Mystery Box, and closed the lid. I would be remaining in Macassar by His appointment, and I could trust the future to Him.

The principal of the Bible School left for furlough, and Ernie Presswood arrived to replace him. Ernie's lovely new bride, Ruth, was appointed to language study.* Ernie and Russell had been schoolmates at Nyack Missionary Training Institute, then were teamed

---

*Prior to the field conference and shortly thereafter, several new missionaries joined our ranks. When hostilities ceased and the smoke of the awful conflagration had lifted, when the sound of the instruments of death were no longer heard in the land—of them, only Ruth Presswood remained. Grace Dittmar, a beautiful character whom we enjoyed immensely and corresponded with after she went to Sumatra, was weakened by the ordeal of her evacuation ahead of the Japanese invaders and succumbed after reaching the States. John Willfinger, a wholesome extrovert, returned to the Sesayap district of East Borneo after the conference. Rather than put the Christian Dyaks in a position where they would be forced to lie if they hid him from the Japanese, John voluntarily surrendered and was bayoneted to death on the island of Tarakan, just off the coast of East Borneo. Fred Jackson, a splendid pilot-cum-missionary, fell victim to the sword of Bushido in Borneo. Helen and Andrew Sande, gentle people, dear personal friends from St. Paul and Nyack days, had been with us in Macassar until after the birth of Baby David Jerome. My last letter from Helen, from the Boelongan district of East Borneo, asked me to have some dresses made for her. I was never able to deliver them. The fury of the sons of the Rising Sun spared neither father nor mother nor child. Over their graves, known and unknown, the finger of God wrote His own triumphant epitaph:

They overcame him
by the blood of the Lamb;
and by the word of their testimony;
and they loved not their lives unto the death.
Revelation 12:11

together in East Borneo. The Presswoods moved in a block from us on the same street. Our friendship flourished. Ruth and I found real rapport.

The missionary staff gathered on an early December Saturday to celebrate the return of the Jaffrays from a brief Philippine holiday. The change had been good for them, and we had much to share with them of the many blessings of the past few days.

The next morning—December 8, 1941—I called Russell and the Jaffrays to join me in the dining room. "Breakfast!" Time for the seven o'clock news; Russell leaned over to turn on the radio. Pearl Harbor had been bombed by the Japanese, December 8, 1941 (December 7 by Washington, D.C., time), a day that would live forever in infamy. America was now a participant in the global conflict.

The reality of war hit home. We began to link the events of the past year together, trying desperately to understand what had brought about such tragedy. The violence portended much more to come. We knew we'd best learn what we could from the past to realistically prepare for the future.

For years Japan had hung in the Pacific Ocean like a giant octopus—her tentacles, her striking forces, concealed. Her nation's leaders chafed under the established ratio in naval strength between the three important naval powers of the world—Great Britain, the United States, and Japan—a ratio of 5:5:3. Japan's demand for equality in naval strength with the United States and Great Britain was not acceptable to the negotiating nations, so in defiance Japan walked out of the second London Naval Conference in 1935. Immediately the militarist leaders in the Japanese administration launched a five-year warship construction program. Two years later, seeking land to relieve the pressure of their burgeoning population and to expand her economy, Japan felt confident enough to spark off the "China Incident," which eventuated in the cruel eight-year Sino-Japanese war. A tentacle reached out and China, already fragmented by civil war, was easy prey.

The United States accepted the Japanese sinking of the U.S. gunboat *Panay* as an error in identification by coastal batteries and bomber pilots. Nevertheless, as a deterrent, in 1940 Congress placed economic sanctions upon the export of potential war materials to Japan. Chemicals, minerals, aircraft components, aviation gas, and lube oil, as well as all scrap iron and steel, became prohibited exports.

About this time I saw a cartoon in an American publication depicting a grinning Japanese soldier standing behind a loaded cannon pointed toward the United States. The caption read, "So you want your scrap iron back?" I cut out the cartoon. Japan was no longer a little-known, remote island empire; she was a close neighbor of the Netherlands East Indies. The response of most to the possibility suggested by the cartoon was, "Ridiculous! They wouldn't dare." They not only dared; they were prepared.

Japan had a navy, an army, and an air force who had been subjected to most rigorous training, men who were disciplined to fight in the tradition of the Samurai and the Bushido. The fatalistic philosophy of the kamikaze enabled the air attack corps of their air force to make suicidal crashes, flying into targets in airplanes containing explosives. There were to be times when I would wonder what had happened to the noble concepts of "benevolence" and "self-control" embodied in the chivalric code of Bushido behavior.

When Japan started her forays into Indochina, President Roosevelt froze all Japanese assets in America. From that moment, historians record that war with Japan was predictable, inevitable, and imminent. At the Imperial Conference presided over by Emperor Hirohito, September 6, 1941, Japan resolved to go to war with Great Britain, the United States, and Holland if their demands were not met by early October. These demands entailed closing the Burma Road, noninterference in China and Indochina, no increase of military forces in the Far East (even in a country's own possessions), and American assistance in obtaining needed raw materials and in establishing good economic relations with Thailand and the Netherlands East Indies. For these concessions, Japan agreed that it would use Indochina as a base of operation against China only, and that as soon as a "just peace" was established in the Far East, she would pull out of Indochina. Japan further agreed to guarantee the neutrality of the Philippines.

Military logistics were never my forte, but when the Premier, Prince Konoye, resigned and War Minister General Tojo invested himself with the premiership, retaining the war ministry and assuming the portfolio of Minister for Home Affairs, it left me no doubt that the militarists were firmly entrenched in the Japanese administration.

Desperately in need of oil and raw materials, Japan was looking south at the Netherlands East Indies, rich in natural resources. Realizing how quickly the sea lanes might be closed, we dispatched as many as possible of our students back to their homes or to their assignments. For those for whom it was impossible to get passage, we established a camp thirty kilometers from the city, with sufficient land for huts and gardens. We knew that if the Japanese were not soon stopped, Macassar's fortifications would be tested. We wanted none of our students to face the already infamous shock troops, should invasion come. These shock troops were the first waves of invading soldiers who, by their cruelty and heinous practices, paralyzed the inhabitants with fear.

It was very reassuring when the BBC news broadcast that a new British battleship, the HMS *Prince of Wales,* had been dispatched with all haste to the Far East, to be joined en route by the veteran battle cruiser HMS *Repulse,* which was somewhere in the Indian Ocean. It was also confirmed that a new aircraft carrier, the *Indomitable,* would be sent to reinforce the British fleet.

Convoys of troop carriers also began to converge on Singapore, reputed to be the strongest naval base in the world. Australian ships, en route to Singapore, disembarked their youthful cargo of soldiers in Macassar. Aware that few of them could speak Indonesian, we went to the city to offer our services as interpreters while they bought souvenirs and trinkets for sweethearts and family "back home." They were so young, so confident, so full of swaggering courage—too young to fall victim to the kamikaze pilots and the sons of Bushido. Such young confidence ought not to be shattered so soon; such young courage ought not to be tried immediately to the breaking point! Many never lived to carry home their gifts of love. Those who returned had grown old, long before their time, in the prison camps of Singapore.

Winston Churchill, in *The Hinge of Fate* (a volume in his *Second World War*), admits that the sinking of the *Prince of Wales* and the *Repulse* was the most direct shock he experienced during the whole course of the war. It was a shock suffered by everyone who had counted on those modern warships to turn back the tide of Japanese victories. Psychologically, the Japanese victory in the

Gulf of Siam was shattering. Through the radio broadcasts describing that victory, we again became poignantly aware of kamikaze pilots. "We fear what we do not understand" became the watchword of the times. We *did not* understand—what kind of fatalistic brainwashing gives birth to suicide pilots? What defense is there against men who dive their planes into enemy ships and the arms of sure death? Our radio gave graphic eyewitness accounts of the kamikaze attacks and the sinking of the ships where they lay in the roads at Singapore, unable to maneuver to safety.

We prayed earnestly for counsel from the mouth of the Lord. Russell was asked to conduct a week of special meetings at the Chinese church in Macassar. The British Vice-Consul, Mr. Lie, was the interpreter. The Lord blessed, and there were those who accepted Christ as their Savior, others who sought a spiritual renewal. God met many hearts and prepared them for the days ahead.

Following Pearl Harbor and the sinking of the British warships, we kept our radios turned to Manila for the newscasts. Little-known islands—mere dots in the vast Pacific—suddenly came to the attention of the whole world as the Japanese attacked, invaded, and occupied them—Guam, December 11; Wake Island, December 23. "Never fear," we were counseled, "the British will stop them at Hong Kong; it's well fortified." But on a bitter Christmas Day, the "impregnable rock" surrendered.

Drunk with victory, the Japanese surged west and south onto the mainland—China, Korea, and French Indochina. In a pioneer maneuver, thrusting from the southern and northern ends of the Philippine Islands, the Japanese backed the defending Filipino and American forces onto the Bataan Peninsula to endure a night of horror out of which few emerged to greet the dawn of the day when peace was signed.

How do you speak courage into the hearts of people when you have nothing but defeat to report? Yet every broadcast from the Manila newscaster held overtones of hope—hope for a miracle that would turn the advancing, victorious hordes of the enemy. I can hear it now and a lump rises in my throat—the last transmission: "I'm here in our building in the center of the city. The bombs are falling all around us. [Background noises of planes and explosions emphasized his statement.] I'm going to have to close down!"

A sob interrupted his delivery. Then all of a sudden, we heard him shout, "Come on, America!"

The awful silence that followed, pregnant with horror, left us sobbing, "Yes, come on, America! Dear God, have mercy!"

January 2, 1942, the Philippines had fallen.

We turned our radio dials to London's BBC and Sydney's ABC to keep abreast of happenings in the European war theater, as well as the Far East. With our finger on the atlas, we traced the rapid advance of the undefeated sons of Japan. They moved forward, making their onslaught on Thailand and Burma, then began the sweep down the Malay Peninsula. The Japanese were jungle fighters who never felt in the least inhibited in their military operations by the Geneva Convention for the treatment of prisoners. The tentacles of Japan had surfaced, had reached out in all directions, and her victims now hung limp in a stranglehold of military might.

While surging forward toward Singapore's back door, Japan had begun its conquest of the Netherlands East Indies. The Dutch armed forces and the Aussies, who had slipped out of Singapore and through the naval net thrown around that island, fought bravely in Sumatra, but the odds were too great. Some escaped to Australia to fight another day; others were incarcerated. Many died.

As soon as the school year ended, Dr. and Mrs. Jaffray, Margaret, and the other single ladies had gone to Benteng Tinggi. They needed rest and respite from the humidity and heat of the coast. The Presswoods, Russell, and I remained in Macassar to purchase basic food supplies and other essential items that would be needed at Benteng Tinggi, war or no war—kerosene, lamps, wicks, and so on. I packed a few personal belongings and linens sufficient for a few months—surely the tide must turn soon. I looked at the trunks full of very beautiful wedding gifts and pulled the door of the storeroom shut, never to see them again. We then joined the others in the mountains.

Five kilometers beyond Benteng Tinggi was Malino, a resort area used by Dutch and other expatriates. There was a steady influx to the area of women and children from government posts throughout the island. Patrol officers remained at their posts. Dutch police were appointed to Malino for the protection of those who sought asylum there.

One by one the islands fell to the Japanese—Sumatra, Borneo, the Lesser Sunda Islands, Bali, Lombok, Sumbawa, Butan, and Muna. Java was under attack. We climbed to the crest of the mountain behind Benteng Tinggi and through binoculars saw numerous ships moving down the Macassar Straits. Distant explosions and billowing black smoke attested that a naval battle was in progress. In their battle-gray coats and camouflage, it was impossible to tell from that distance whose ships they were. Not until years later did I learn that what we saw was part of the terrible Battle of the Macassar Straits, in which so many Allied ships and men were lost. We knew it was inevitable that the island of Celebes be invaded.

On a Wednesday, a Dutch policeman came to Benteng Tinggi from Malino to inform us that they had a ship lying at anchor on the south coast. They wanted to evacuate all foreigners and all Dutch women and children who wished to go. A truck would call for us on Friday, so we should be ready.

As we gathered for prayer, Dr. Jaffray said, "I want to counsel you not to discuss this decision that must be made with each other—not even husband and wife. Go to your knees and say, 'Lord, what do You want me to do? Shall I go or shall I stay?' This is extremely vital, because then no matter what happens in the months or possibly years that lie ahead, you will know that you are exactly where God wants you to be. If He leads you to leave, you'll never feel that you were a coward and fled If you are led to stay, no matter what happens you can look up and say, 'Lord, you intended for me to be right here.' " We earnestly sought guidance.

When the truck arrived on Friday, there was not a person among us who felt led to leave. As Dr. Jaffray had said, "God does not work in confusion, a wife against a husband or vice versa, in a matter that concerns both of you. This is but a confirmation to your hearts of His directive."

Some three days later, it was reported that the ship had been torpedoed and sunk. There were no known survivors. Then I knew why God had said, "Don't go." It is imperative that we know the voice of the Shepherd and learn to follow Him when He speaks. We must be obedient, no matter what He says to us; it may even mean our life.

The harbor at Macassar was heavily fortified, but the enemy, having been well informed by our former barber and storekeepers, came in the back door, landing on the unprotected beach at Borombong on February 8, without a shot being fired. The shock troops swept on into Macassar, killing right and left as they went. They found that everything in the city that might aid the Japanese war effort had been blown up and burned. The Dutch army had retreated into the mountains but soon surrendered.

The fear of the invaders pervaded the island. Planes strafed Malino with machine-gun fire, but Benteng Tinggi was spared. Who but those who had experienced it could really understand the apprehension of those days, waiting for the captors who would surely come? Reports filtering in were not reassuring. How would we be treated, or mistreated? Were our very lives to be spilled because of the hateful whim of some enemy soldier? Would there be torture? Could we endure? With these and a multitude of other questions, we psychologically weighed our preparedness to face the unknown—a confrontation with the enemy. Always we came back to that which brought quietness: God intended for us to be here. This was His appointed place for us at this time.

Connected to the Jaffrays' house by a covered walkway was a pavilion containing two bedrooms, a bathroom (dipper), and a kitchen. These bedrooms served as guestrooms. The Presswoods occupied one, and Russell and I the other. Our first night at Benteng Tinggi, I was awakened by an outcry and loud talking, which shot me up in bed, convinced that all the armies of the Rising Sun had suddenly arrived. I was covered with cold sweat.

Russell then awoke and informed me that it was only Ernie Presswood talking in his sleep. "You get used to it," he admitted casually, with Ernie still moaning and babbling on the other side of the wall. "When we were in Borneo, we lived with an old cook who snored something fierce. The hut we stayed in was a crude structure with no ceiling, just bare roof. Our first night there, Ernie made so much noise in his sleep that he even roused the old cook. The next morning the old man said to me, *Tuan, bukan main, orang itu melawan saja!* "Sir, it's no joke, that man gave me competition!"

This afforded some comic relief to our tension. Ruth informed me the next day that Ernie had been preaching in Dyak. Echoing Russell's words, she said, "You get used to it."

The conference property was not visible from the main road, but we daily heard the enemy trucks grinding their gears as they passed on the way to and from Malino. Because we were certain they would by now have noticed the lane that turned into Benteng Tinggi, we wondered how soon it would be before the enemy came to our doors.

On March 5, while working in the garden, I was attracted by a noise in the yard and looked up to see a Japanese soldier wearing black tennis shoes with the big toe separate (for climbing ropes on ships) rounding the corner of the house. The Japanese had come. The soldier pointed his gun, with fixed bayonet, at me, motioning me toward Miss Marsh's house. As I was being propelled reluctantly forward, Russell, the Jaffrays, and more soldiers joined me and my escort.

We were herded into Miss Marsh's living room to join the Presswoods and the three single women. While we stood at attention (a soldier with a gun pointed at your back tends to make you do that), the two officers slouched in the chairs. The soldiers' uniforms were tattered and faded, their faces hidden behind unkempt, matted beards. They were filthy from who knows how many days of island-hopping and marching through the jungle. Their headgear were caps with a bill in front and a flap in back to protect their necks from the sun. These were the infamous "shock troops"—a chill went through me.

The commanding officer announced from his very relaxed position that we were prisoners of the Imperial Japanese Army. Russell was standing in his customary way, with his hands in front of him, the palm of one hand resting on the back of the other. It would have been impossible for him to have been holding anything in his hands, but the posture nevertheless infuriated one of the officers who snapped a brisk command in Japanese. A soldier strode forward, raised his sheathed bayonet, and began to beat Russell's hands again and again. Russell dared not resist. I was appalled. Finally Ernie could stand it no longer, so he said, "Russell, they want you to put your hands down at your sides."

In helpless anger I thought, "You dirty rats. If you had just said so, but. . . . " Ernie knew what they wanted, for when he had first met the soldiers, he had raised his hands above his head. This infuriated the officer, and with his sheathed sword he beat Ernie's arms unmercifully. Realizing that his gesture of surrender incensed

the man, Ernie dropped his hands to his sides and the flailing
ceased. This senseless maliciousness had its desired effect; we
were greatly subdued. However, Russell said later to us, "They hurt
my pride more than my hands."

We were asked to state our nationalities. When Margaret Kemp,
Philoma Seely, Russell, and I identified ourselves as Americans, the
officer announced, in a tone of voice meant to humiliate, that the
whole American navy had been sunk, and we were their prisoners.
Then he came to Dr. Jaffray, who replied that he was from Canada.
Like the rest of us, he was terribly nervous and stuttered over his
answer. The interrogator looked at him and questioned, "Kah-nah'-
na-da? Where's Kah-nah'-na-da?" He obviously had never heard of
Kah-nah'-na-da before, so we felt less worried for the Jaffrays and
the Presswoods. Miss Marsh was from England, and apparently he
had never heard of that country either.

While this investigation was being conducted, Ruth Presswood
tried surreptitiously to inch her way around a chair to Ernie's side.
A brisk command was given, and a soldier propelled her back to
her place with the butt of his gun.

Finally making preparations to go, they again impressed upon
us that we were prisoners of the Imperial Japanese Army; we were
to have contact with no one outside the premises, nor were we to
leave the conference grounds. If we did, the penalty would be very
severe—we would be shot!

By March 8, 1942, just three months from that Sunday when
we listened to the news that Pearl Harbor had been bombed, the
rape of the Netherlands East Indies was complete. The small Dutch
army garrisoned in the islands and the tiny Dutch navy and air
force were helpless before the onrush of landing forces that flowed
in a seemingly unending wave from Japanese transports.

The days went by, tension eased, and we settled into a loose
pattern of cooking, eating, Bible study, prayer, gardening, reading,
and walking the perimeter of our property. I was in the house one
morning when I heard a truck approaching in the distance. It
stopped—"The soldiers must be across the gully at Miss Marsh's
house"—then it rumbled into the Jaffray yard. Japanese soldiers
and an officer entered the house. "We're taking the men," they
announced. "Get some clothes together for them. No suitcases.
Quickly!" Alarm filled me.

Running out to the pavilion, I found a pillowcase and put into it Russell's Bible, a notebook, a pen, shaving gear, clothes, and other things I thought he would need. Hurrying back into the house in search of him, I met an officer coming from Dr. Jaffray's room. "What's wrong with that old man?" he demanded of me. There were many things, and it took quite a while to enumerate them—he had diabetes, so he had to be on a very strict sugar-free diet; he had been in a coma not too long before we had come to Benteng Tinggi; he had kidney trouble, a heart ailment, involuntary physical shaking that affected his right hand and arm in particular, and. . . . Before I could say anything more, he shut me up with a wave of his hand and snapped, "Go in and tell that old man he doesn't have to go. If he needs all that medicine, he's not going to live very long anyway."

I dropped the pillowcase into a chair and ran to the door of Dr. Jaffray's bedroom. I caught a glimpse of him putting things into his black satchel (which had belonged to his senator father). "They say you don't have to go, Dr. Jaffray." I retrieved the pillowcase, then dashed out into the yard, my eyes searching for Russell. Where was he? What had they done to him? Then I saw him; he was already in the truck with the other POWs. He was standing in the back near the tailgate. I was terrified for him. Every unreasonable fear told me that he was being taken away to be executed. Why had the officers said that Dr. Jaffray would die *anyway?* All the other separations we had endured gladly, for those had entailed the cause of Christ's Kingdom, but this—this was different; the thought of this separation was excruciating.

I handed Russell the pillowcase and looked into the face that had become so dear to me. A cry of protest, of fear, strangled itself in my throat. "You sadists, you didn't even let me say goodbye!" I swallowed hard and clenched my fists. "You'll not have the satisfaction of seeing me cry." The driver started the engine. Russell leaned over the tailgate and very quietly said, "Remember one thing, dear: God said that He would never leave us nor forsake us." The truck started with a jerk and disappeared down the road. I never saw him again.

It was Friday the 13th—the 13th of March, 1942. I have a bit o' Irish in me, but I'm not at all superstitious. I truly believe Romans 8:28: "All things work together for good." We don't find it difficult to repeat the verse and say we believe it when *all* things are going well. But when we find ourselves going through deep waters, confronted by trials we don't understand, can we then say, "I believe that *all* things work together for good?" At that moment I couldn't.

Everything had happened so fast and without the slightest warning. Russell had said, "He will never leave us nor forsake us." No? What about now, Lord? This was one of the times when I thought God had left me, that He had forsaken me. I was to discover, however, that when I took my eyes off the circumstances that were overwhelming me, over which I had no control, and looked up, my Lord was there, standing on the parapet of heaven looking down. Deep in my heart He whispered, "I'm here. Even when you don't see Me, I'm here. Never for a moment are you out of My sight."

Was this the Lord—this quietness that was beginning to seep into the crevices of my hurt—or was it nothing more than shock to find myself abandoned to the caprice of the enemy? I turned toward the house to find Dr. Jaffray and Margaret waiting for me. I threw my arms around them. "Oh, Margaret, thank God they didn't take your dad. And Dr. Jaffray, what were you doing in the bedroom when the officer was with you?" It was puzzling that any consideration had been shown for Dr. Jaffray's condition—quite out of character for the Japanese we had met previously.

"I was just putting my eau de cologne in my father's satchel. I figured we would be taken down to the coast. You know how refreshing I find eau de cologne in the humidity, so I thought I would take some along."

Eau de cologne! And the officer had thought it was medicine!

"Oh, Dr. Jaffray, isn't that beautiful? Isn't that just hilarious? The Lord really tricked that officer into thinking all that eau de cologne was medicine." As I told them about how the officer had cross-examined me concerning Dr. Jaffray's physical condition, we started to laugh. We laughed until we were weak. Nerves? Perhaps partly, and "a merry heart doeth good like a medicine" (Proverbs 17:22), but beyond that we realized that God, who knows the end from the beginning, knew how Dr. Jaffray came to have so many bottles of eau de cologne and how He would use it to confuse a Japanese officer so that he decided to leave him with us—seven women who needed the physical presence of that man of God.

When the sunset and night came on, the full import of my loss hit me again: Russell was gone. We had put another single bed in Margaret's room, and I moved into the house rather than remain alone in the pavilion. We sat down to have prayer; Dr. Jaffray opened the *Daily Light* to read—March 13. For me in my need, the Lord had directed in the arrangement of the verses. The Psalmist echoed the cry of my heart:

O my God, my soul is cast down within me.

Thou wilt keep him in perfect peace, whose mind is stayed on Thee: because he trusteth in Thee. Trust ye in the Lord forever: for in the Lord JEHOVAH is everlasting strength.

Cast thy burden upon the Lord, and He shall sustain thee. He hath not despised nor abhorred the affliction of the afflicted; neither hath He hid His face from him; but when he cried unto Him, He heard. Is any among you afflicted? Let him pray.

Let not your heart be troubled, neither let it be afraid. . . . Lo, I am with you always.

"Let him pray," the verse said. Mounting the steps into His presence, I prayed and He came to me with the gift of remembrance of a little girl saying, "Lord, I'd go anywhere with You, no matter what it cost." Was that but an expression of childish enthusiasm resulting from an emotion-packed presentation of the mission field?

"I meant it then, my Lord, to the level of my understanding. With greater understanding I confirm to You tonight, it is still *anywhere*—I leave the costing to You." He took my hand, and together we walked into a future yet unknown. But from that moment, the sting was gone from the wound.

When we went to Benteng Tinggi with a minimum of supplies, we had no idea that our internment there would stretch into a year, followed by three more years in various camps and prisons. Consequently, in order to survive, it was necessary to be ingenious about our food sources. The first time one of the mountain women, whom I had led to the Lord, came with a half coconut shell filled with goodies that had been fried in coconut oil, I accepted them gladly. I tried one and found it delicious, then remembered that I had seen the native children the day before out on the mountainside collecting insects. Could these be . . . ? "Rassing, are these flying ants?"

"I'm glad you like them, *Njonja.* I'll get you some more."

We also ate various weeds and jungle plants. Ferns are truly delicious; I had learned to eat them from the Indonesians. There was one weed that had a tiny blue flower. The Dutch called it *hemelse blauwe oogen* "heavenly blue eyes." We renamed it "watercress," feeling that our nomenclature made it taste better. Wherever the Japanese went, they planted Spanish needles. We boiled them like vegetable greens. The finished product was something like boiled okra, very gooey and slimy. From Rassing we obtained some seeds and started a garden.

We never ceased to be grateful to our spiritual family from the Macassar Tabernacle, who, learning of our plight, risked their lives to bring food to us lest we starve to death before our garden matured. Without their help we might have done just that, for no rations were given us by the Japanese. In the face of extreme punishment if apprehended, our friends came over the mountain behind the property, bringing what rations they could afford and carry. They also brought whatever news they could glean about Russell and Ernie. War news had a Japanese flavor—the whole of the Far East had surrendered to them.

Our friends had kept our house in Macassar under close surveillance. Soon after the surrender of the city, Japanese soldiers had forced entry into our house and storeroom. My trunks of wedding gifts were dragged into the yard and the locks forced. Whatever the soldiers didn't want they scattered about the driveway. Our new refrigerator was carried outside by several Japanese who, one-two-three, threw it so hard that when it crashed on the gravel, the door flew open and irreparable damage was done to it.

Bit by bit treasured keepsakes and souvenirs were being wrested from me. I was being taught to live so that my most treasured mementos took the form of beautiful memories stored in the file of my heart, where moth and rust—and Japanese soldiers—could not corrupt or destroy. Whatever He allowed us of creature comforts, we were free to enjoy. If He chose to allow others to take them from us, that was His prerogative.

Japanese soldiers visited at regular intervals as they passed Benteng Tinggi on their trips to and from Malino. We were in constant dread of their harassment and, apart from God, there was no protection against the power-hungry who might choose to avenge himself on us as symbols of the hated American enemy.

One day after the soldiers were gone, Margaret and I ran into Dr. Jaffray's bedroom to see if they had taken his watch. Ordinarily it hung on a velvet holder set on the table at the foot of his bed. It was gone. "Dr. Jaffray, did they get your watch?"

He answered by lifting his pillow, retrieving his flashlight and the watch, and returning them to their usual places.

Time after time when the marauding soldiers had left the property, Dr. Jaffray lifted his pillow and took from under it his watch and flashlight. "Whenever I hear the Japanese have been sighted, I put them on the bed and pull the pillow down over them. I remind the Lord that neither of them means a thing to the Japanese, but they do to me. The watch was my father's. The Lord knows I need it and the flashlight; they don't."

His confident expectancy that the watch and flashlight would always be there thrilled me—a practical lesson in faith. It was not until some time after the war that I realized the flashlight had burned the entire time we were under house arrest in the mountains. Although used every night, the batteries were never replaced, nor could they have been, for we had no access to new ones—a miracle of God's provision.

Lumber that has not been properly dried is not the best material with which to build a house, but this was the kind that had been used in the Jaffrays' house. As the lumber shrank, openings appeared between the planks of the single-walled house. Rats could enter without having to expend energy gnawing their way in. They were looking for food and a dry nesting place as the rains had already started—where better than with the missionaries? We had no rice or other food to share, and we couldn't afford to have them gnawing the leather of our shoes or using our clothes, paper, or books for lining their nests, so Margaret and I declared all-out war on them. Afternoons while it was still light enough to see, we started in the bedrooms at the back of the house running the rats down the hall and into the kitchen where the doors fit tightly and the walls had no holes. Then each of us, armed with a broom, fought the rats until we had killed all we had coralled. We squealed and yelled when the rats ran up the wall and jumped on us. It was a nasty business, and I especially disliked the cleanup job that was necessary after the kill.

One night my rest was disturbed by what I thought were rats. I could hear them moving around the livingroom, in the dining

room, and along the halls. I tried to ignore them with a promise:
"I'll get you tomorrow." But when I heard a book fall to the floor,
that did it. I jumped up, giving Margaret's bed a shake, and called,
"Margaret, grab your robe. We'll light the lamps and have another
go at the rats. I've been hearing them from one end of the house
to the other."

A hall ran the full length of the house. Our bedrooms and the
kitchen opened off it with the living/diningroom on one end and
a bathroom on the other. When I pulled the bedroom door open,
in the dim light of a little night lamp I saw someone swish past
me. I thought it was Dr. Jaffray and was perplexed at his strange
behavior in the middle of the night. When I stepped out into the
hall to get a better look, I found myself face to face with a Boegis
bandit.

He was wearing a black sarong that he swooped up over his
shoulder to free his machete. With one fluid movement, the knife
was extricated from his belt and held up in striking position. I'm
really quite a coward, and why I rushed at him, I have no idea.
Perhaps it was the element of surprise, but he was a bigger coward
than I, for he turned and fled down the hall, through the bathroom, across the porch, and down over the mountainside with me
hot on his heels—until I saw others emerge from the jungle. He
yelled something in their language, and together they fled. I stopped
dead. "Lord," I whispered, "what a stupid thing for me to do!"

Immediately He answered, "The angel of the Lord encampeth
round about them that fear Him, and delivereth them."

I went back to the house and reached to pull shut the door
that was standing wide open. There was no doorknob, no lock or
key; instead, I pulled the door toward me to find a porthole carved
in the door by the intruder with his long knife.

Dr. and Mrs. Jaffray were now awake and had joined Margaret.
"What happened?" Dr. Jaffray asked, much shaken by the noise and
sight of the damaged door.

"We had bandits! They must have been here for hours. I thought
they were rats!" A tour of inspection proved that tablecloths and
other linens were missing, books had been pulled from the shelves
and searched, probably by someone expecting to find money hidden in them—and I had thought they were rats! They curiously
had taken panes out of Dr. Jaffray's bedroom windows and laid

them carefully on the porch so that we might replace them. None of the noise had awakened Dr. Jaffray. I found a board and nailed the door shut.

From that night on, we slept with a club at the foot of our bed and a small milk can–squawker under our pillow, but we never had to use them. We heard bandits return several nights after that, but they never again entered the house. It wasn't until after the war that I learned why. I had suspected the Jaffrays' gardener; he was Boegis, and he knew the layout of the house. When I asked him why they had never entered the house again, he answered incredulously, "Because of those people you had there—those people in white who stood about the house." The Lord had put His angels around us. He had delivered.

One day the Japanese stopped at Benteng Tinggi with a Dutch nurse, a friend of mine, who was being transferred from a prison in Macassar to Malino. They allowed the nurse to get out of the car and come inside. As they left the house to continue on to the Dutch camp, the nurse slipped a fountain pen into Ruth's hand. Ruth recognized it as Ernie's and wondered why he would be sending it to her. As soon as the truck left the property, with hands shaking with excitement she began to take the pen apart. Sure enough, rolled into a tight cylinder and secreted in the tiny ink tube was a letter. We were all feverish with excitement by this time. Ernie wrote that he and Russell were imprisoned in the police barracks in Macassar but expected to be moved soon. They had sufficient food and were well. Ruth shared the letter with us all until we knew it by heart. Like thirsty people returning to the fountain, Ruth read and reread the letter to us. We discussed it and pondered it, not wishing to miss some hidden message. That was an exciting day—at least we knew where they were, and that they were well.

My turn came. The bearer of the letter was a Japanese naval admiral. He spoke excellent English and had spent some time in America. He had been quite impressed by Annapolis. He was dressed in an immaculate white dress uniform and was very polite and friendly. When he told me how much he had enjoyed visiting with Russell and handed me a letter from him, I felt he was not being sneaky or simply polite. He seemed genuinely concerned about our welfare—did we have enough to eat? Could he get anything for us? There was none of the rudeness or bravado we had previously experienced. He didn't even mention the war.

As soon as he was gone, we devoured my letter. It was similar in content to Ernie's. They were getting rice to eat. Rumors had it that they were to be moved to Pare Pare, a coastal town approximately 100 kilometers north. They had heard that military POWs were to be imprisoned in the police barracks after the civilians were evacuated. There had been women in their prison who were to be transferred to Malino. "I'm so concerned about you. I wish 1001 x's I had taken you away from here." This was the P.S. that closed his letter.

I could understand his concern. This was war. Many immoral Japanese soldiers took advantage of women who couldn't protect themselves or be protected by their husbands. We prayed for peace of mind and heart for both Russell and Ernie and an assurance that we were safe in God's protective custody.

Even though a Japanese officer had brought my letter, Russell had written freely about their expected move to Pare Pare. Fearful lest one of the booty-hunting soldiers discover my missive, I hid it on top of the plasterboard that served as a ceiling. I was afraid they might punish Russell for contacting me. The admiral might by then have shipped out of Macassar; I had no way of proving how I had received the letter.

One day, needing to reread Russell's letter, to see his handwriting again, I clambered up to remove it from its hiding place. To my dismay, it was gone. I felt around—no letter. I was terrified. Had one of the looters found it? I yelled for Margaret to come hold a chair on top of a stepstool so that I could climb up into the dark unfinished attic. I was very careful to crawl around on the crossbeams, feeling with my hands over every inch of the ceiling. Perspiration started to trickle down my cheeks. "O God, what happened to my letter? If it's here, help me find it." My hand came down on a crumple of paper sheltering some baby rats. A mother rat had shredded my letter to make a nest for her litter. I eased my way back across the crossbeams, firmly clutching the pieces of Russell's letter and the bare pink bodies. They had to be destroyed separately, but I felt great relief when the last shred of paper disappeared down the toilet. No one would ever find it now, and I knew the letter by heart, so it wasn't really lost to me. I thanked the Lord and promised that if I ever got another one, I'd destroy it as soon as I had committed it to memory.

Margaret was extremely careful with the foods she prepared for her father, because of Dr. Jaffray's diabetic condition. We had stockpiled saccharin to substitute in every recipe that called for sugar. One afternoon Dr. Jaffray, Margaret, and I walked down the road as far as we were allowed to go. Returning and feeling somewhat exhausted, Dr. Jaffray sank into his chair and suggested that we have a cup of tea before we started the evening meal. Margaret prepared tea and set the tray on a small table near her father. He helped himself to milk, and, instead of taking saccharin, he picked up the sugar bowl and spooned one, two, three teaspoons of sugar into his tea. I couldn't believe my eyes, and Margaret was horrified. "Oh, Daddy," she pleaded, "please don't do that. You know you aren't supposed to have sugar."

Understanding her strong feelings prompted by love and the still-fresh-in-her-mind memory of his most recent coma, he tried to reassure her. "Muggie, I'm healed. I need this sugar for strength." He continued to use sugar, and when she begged that he not use so much, he patted her hand, saying, "It's all right, Muggie. I'm healed."

Not too many days later, Lexje Kandou, the young son of the Macassar pastor, arrived at Benteng Tinggi with food for us. "Lexje," I said, "do you think there is any way you could get a urine specimen to the Dutch doctor in Malino? For several days now Dr. Jaffray has been using sugar. He believes the Lord has healed him, but we would like the doctor's confirmation. We don't want you to get in trouble because of this. What do you think?"

"You give me the specimen, and I'll get it to the doctor, *Njonja*. I'll need some of the vegetables back, so that the Japanese will think I've brought them into Malino for the market." We did as he asked, then commended him to the Lord for protection. This was a dangerous undertaking, but he was a very brave teenager. A few days later, a friendly native brought a letter from the Dutch doctor saying that he had examined the urine specimen and found not a trace of sugar. Dr. Jaffray had indeed been healed.

"You see, Muggie, the Lord had healed me. I knew He had." We had a time of praising and thanking the Lord. He was preparing Dr. Jaffray for a time when there would be no saccharin, only scant

rations of sugar. The Lord is very good to those who put their
trust in Him.

We missionaries came to know one another well because of
the fellowship of our sufferings and victories. Despite the separation and difficulties, our imprisonment was a time of private spiritual tutelage and lasting blessing for me. I think often on that
which Dr. Jaffray shared concerning his childhood, his home, his
Christian mother, his father, whom he greatly admired, his Bible
School days, and the years in South China.

Our gardens were maturing and needed weeding. This was in
our morning agenda, along with lessons for Bible School in anticipation of the day when war would cease and we could return to
our work. Miss Lilian took dictation from Dr. Jaffray in Chinese for
the *Bible Magazine*.

Mrs. Jaffray did beautiful handsewing. She had profited from
her years with her beloved Chinese. She mended everything that
was in disrepair. With knitting, tatting, embroidery work, afternoon
Bible study, prayer meetings, and our other pursuits, the days
passed. Evenings, using candles, Margaret and I read our way
through an unabridged set of Charles Dickens.

I had spent several hours one morning weeding, and when I
stood up to go into the house for lunch, I felt lightheaded and
quite ill. I had been wearing a hat, but the sun was exceptionally
hot. I knew I had to lie down. When I became delirious and
feverish, Margaret ran to get Ruth, who came running with all the
paraphernalia of a well-trained RN. My fever was high. I suddenly
became conscious of Ruth and Margaret on either side of the bed,
others at the door, and Dr. Jaffray gripping the rail at the foot of
the bed. He began to pray; of this I was very much aware: "Lord,
you know how terrible it is for all of us when the soldiers come
here. Keep them away while Darlene is sick. Don't let them come
on the property again during this time."

My fever peaked every afternoon, and I was not allowed out of
bed. I really didn't feel like arguing. Where had my strength gone
in that one hot morning in the garden? Was it sunstroke? They sent
word by Lexje to the doctor in Malino. As before, Lexje carried
vegetables to sell to the people at Malino as a legitimate excuse
for being on the road. (Lexje also shared the news of us missionaries with our friends at the Tabernacle in Macassar.) Through

Lexje the doctor said I should remain in bed. He thought it was a type of malaria, though no blood test was taken.

I spent six weeks in bed before I was free of fever. The first day I got out of bed and dressed, the Japanese came. God had wonderfully answered Dr. Jaffray's prayer.

We received some disturbing news through Lexje. Our friends, who kept regular surveillance of the prison, had seen Ernie Presswood being taken to a hospital ship anchored in the harbor. Was he being treated for some illness, or did he need surgery? Was he to be taken away? Poor Ruth, those were difficult days for her, but at last a clarification reached us: he was suffering from a very serious case of dysentery and had been taken to the hospital ship for treatment. However, by the time the good report reached us, Ernie was back in the police barracks with Russell. Then shortly thereafter it was confirmed—the civilian men had been transferred to Pare Pare.

As if to make up for lost time, Benteng Tinggi became a regular stop for officers escorting women, sometimes children, to Malino. Most of the women were nervous and apprehensive. Some of the women were from countries belonging to the Axis block of nations. Even though they had been allowed to remain in their homes, their movements were restricted. Most of the German women had immediately identified with their Dutch husbands and friends. Their sympathies were not with the Japanese. There was, on the part of many of the Dutch, a certain amount of mistrust and reserve toward the Germans, who had immediately displayed their large swastika and sat under its shelter, while the Dutch women and children were restricted to the overcrowded, unsanitary conditions and inconveniences of Malino or Macassar prisons.

Apparently all women and children were being assembled in one center. On one occasion, we could hear a woman arguing with the Japanese before they got out of the car. She definitely didn't want to come into our house, but the officers were adamant. We could see she had been crying.

"There's no reason why I should have to leave Macassar. I wanted to stay there, but they wouldn't let me!" Tears of anger and frustration spilled over.

Diam! "Shut up!" The Japanese officer's order was not effective, for before they were through the door, she was badgering the officers again to take her back to Macassar. Now what was the big attraction in Macassar? I wondered. Perhaps a loved one in prison she wanted to remain close to?

She walked again onto the stage of my life at a later date, playing the role of a sinister character involved in fraternizing with the enemy. But many traumatic events and changes were to transpire before that day.

Just days later I read the verse, " . . . Your old men shall dream dreams, your young men shall see visions" (Joel 2:28). I closed the book of Joel and arose from my knees. Walking into the hall, I saw before me a partial fulfillment of that prophecy. There before me sat the old man dreaming his dreams. His eyes were closed, but I knew he wasn't sleeping. One hand rested on an open atlas, the other on the arm of the wing chair that had belonged to his father. I knew that by faith he and his Lord were moving down the great chain of islands known as the Netherlands East Indies.

Sensing my presence, Dr. Jaffray looked up and smiled—the smile of one who had had sweet communion with the Lord. I was sure that God had acquiesced to all Dr. Jaffray's proposals concerning reaching the lost, for this, too, was the great burden of the heart of God. How often I had heard Dr. Jaffray remind the Lord of the verse, "Concerning the work of My hands command ye Me" (Isaiah 45:11).

I looked down at the very familiar atlas. Had we not traced the rapid advance of the Japanese on its pages? What other places had fallen under their domination since that fateful day when the island of Celebes had been overrun? His mind was full of warfare, too, but not the same warfare that dominated my thoughts. I knelt beside the chair and listened to his dream.

"Lassie, this is our task. These are the areas we must enter when this war is over." When *this* war is over? It was but beginning. How much more of its fears and anxieties, separations and grim tales of death must we experience before it was over? I suddenly saw Dr. Jaffray as I had not seen him before—old enough to dream dreams, young enough to see visions. To Dr. Jaffray, our experiences were but passing events that never altered God's program of reaching the unreached; never could they mar the old man's dream!

With steady hand and the voice of one assured of victory, he traced upon the map our coming campaign: the Natuna and Anambas Islands; Sumatra—ferreting out and mopping up those pockets of satanic resistance in the central and southern districts; the final liberation of the Punans of Borneo's hinterland; Bali, firmly held in the grip of the enemy, would be freed, its iron gates yielding under the onslaught of faith and prayer. He paused to give praise to our Commander in Chief for spiritual battles fought and won in some of the smaller islands, then moved on to Misool, the Isle of Demons, the Bird's Head of New Guinea, the Wissel Lakes area, the Zwart and Memberamo River Valleys, down either side of the Carstensz backbone—and at last his finger came to rest over the Grand Valley of the Baliem.

"This, Lassie, is our task. Listen—do you hear it?—the sound of a going in the tops of the mulberry trees. It is the noise of the marching army of young men and women whom God is preparing for the day of spiritual battle and occupation of these areas."

I realized how little I knew of what makes a true missionary statesman; of a faith that never staggers at the promise of God, no matter how incredible to the natural man its fulfillment seems; of a trust in the Unchanging One, Who keeps the heart at rest and unperturbed in a changing world; of a burning love that counts not life dear unto itself, but is expendable for God; and of a vision that is never dimmed. Were not these the qualities that characterized pioneers down through the centuries? Were not these the elements that gave the drive, the impetus that launched the missionary campaign of the Netherlands East Indies in 1929, the year of depression when others were retrenching on all sides? Were not these also the characteristics of the godly men and women in the homeland who knew a God whose supply is not modified by the world's economic situation?

Here beside me was the man who had spied out the land and was with the first wave of troops to go ashore in Macassar to stake a claim for God. Once again the world was enveloped in sorrow and difficulties, but these dark days of war were to Dr. Robert Alexander Jaffray, the great missionary general, but days of retreat in which to plan the strategy of yet greater conquests.

I dropped my head on the arm of the chair and found that there were tears on my cheeks. That afternoon I reminded God

that I was available, and never would I call my task common or
mundane if it were a part of the culmination of the old man's
dream, for that afternoon I had seen a vision of the unfinished
task.

CHAPTER FOUR

All women and children, evidently, were to be moved to Malino. Knowing that we, too, could be uprooted, we packed boxes of essential linens and household items and left them by our front doors in readiness. We had only to put a few personal articles in our bags and we would be ready to move at a moment's notice. We had learned that the Japanese soldiers were an impatient people. When they said, *Pigi!* "Go!" you had better be ready to *pigi!*

We had read correctly the handwriting on the wall. The Japanese would not be happy until we were under constant surveillance. In December of 1942, army officers arrived in a truck. Without a greeting of any kind, they made a tour of the house, flicking this and that with their canes and making comments to one another in Japanese. The officer in charge flopped into Dr. Jaffray's chair and, taking in the whole of the house with a disdainful sweep of his hand, announced, "You live much luxury! We take you somewhere intern you."

I glanced at the single-walled wood-frame house, the simple furnishings, thought of our battle with rats and the struggle to stay alive—no thanks to them, for we had yet to receive our first grain of rice or rations of any kind from them. Luxury? We lived "much luxury"?

We went to our bedrooms to collect our personal luggage. I put my Bible, diary, Bride's Book, and hymnal in Russell's briefcase, then added my housecoat and as many dresses and articles of lingerie as it would hold. Our furlough was already four months overdue and my dresses were well worn, with the exception of one. It was a green-and-white-checked seersucker. The skirt was a full circle. It had been sent me as a birthday gift the year before by friends from Boone, Iowa. It had pleased me so much that I had been saving it for the trip back to the States at furlough time.

Margaret and I placed the boxes and other luggage in the truck, helped Dr. and Mrs. Jaffray and the other ladies onto the truck bed, then crawled aboard to hang on for dear life lest we be thrown out during the starts and stops and when the driver changed gears on the steep grades. Most truck drivers pride themselves on being able to shift from one gear to another with at least a modicum of finesse. Not this one. However, he may have been trying to make it difficult for us to stay on our feet. We arrived in the late afternoon at the native market area and were ordered out of the truck. As we unloaded our baggage, the commanding officer pointed out a small house across the valley from Malino. It was just visible through the trees. Then he said, "These men will carry your baggage," and with a sweep of his cane he indicated a number of natives who had gathered to watch the proceedings. With that he left. I was amazed and grateful that we were allowed help but wondered if we would be able to cope with them—all eighteen were Boegis, a larcenous, lecherous-looking lot. I felt the hair rise on the back of my neck, but in an emergency it's amazing how one can rise to the occasion. This was an emergency. To survive we needed every box, parcel, and bag we had.

I suggested that Dr. Jaffray go on ahead with the others, for it would soon be dark. Each one carried his or her small bag of personal items. If it hadn't been so frightening, it would have been hilarious. Dr. Jaffray led the entourage, carrying his black satchel of "eau de cologne–medicines" in his left hand, his cane in his right. Mrs. Jaffray followed, wearing a black velvet cape lined in white satin—for which she thanked the Lord every time she put it on. She looked very regal with her beautiful white hair as she stepped onto the path bearing in her arms a magnificent, large, silver soup tureen, formerly owned by Dr. Jaffray's favorite aunt. Then in single file the others followed after, Miss Seely bringing up the rear.

Ruth and I, as the youngest, volunteered to follow with the carriers and the rest of the baggage. I wasn't about to let those Boegis men know how scared I was. We checked their loads, lined them up in single file, then ordered them to move out with Ruth at the head of the column and me positioned at the rear. I had agreed with Ruth that as soon as we hit the jungle, where it would be fairly dark, we would start circling them. I gave Ruth a signal, then counting in my most martial tones—*Satu! Dua! Tiga! Empat!*—I marched to the front of the line with Ruth circling to the rear doing the same thing. Reaching the first man in the line, I shouted, *Delapan belas!*—"Eighteen!" As soon as I heard Ruth yell "Eighteen!" I'd retreat down the other side, counting the men off again, and Ruth would advance toward the front. We continued this until we finally emerged from the jungle on the other side of the valley. Reaching the small house indicated by the officer, we found the others waiting for us. They had lighted a small kerosene lamp. I ordered the carriers to sit down while I examined each piece of luggage to make sure none of them had been rifled, then handed the boxes and parcels into the house. As soon as the last piece was safe inside, I told the carriers to return to Malino and report to the Japanese officer. When the last man disappeared outside the circle of light, I went inside.

After my previous encounter with the Boegis bandit at Benteng Tinggi, I had a healthy respect for the Boegis and their machetes. These men had all been armed. God had graciously protected us. As I thought of our trip through that stretch of dark jungle, and how easily we could have been killed and all the luggage lost, I felt overwhelmed. I began to retch. Mrs. Jaffray knew I wouldn't make it to the door, so she grabbed the nearest container, her beautiful silver soup tureen, and stuck it under my chin. The levity of the situation struck us and, with the blessed gift of laughter, the tension eased.

We decided that we should bed down wherever we found a spot. It would be easier to arrange things in the daylight. I slept on the table that night, and for several nights following, until I could make myself a bench.

Our house was very small—two tiny rooms, a narrow hall, and a small cubicle for a bathroom. We would have to cook over an open-wood fire under a two-person-sized lean-to on the east side of the house. Ruth and I slept on benches in the room that was to function as our dining room, sitting room, study, and by night our bedroom. Dr. Jaffray slept on a narrow bench in the hall, and the others slept on crude wooden beds arranged dormitory fashion in the second room. It was all terribly crowded. Maybe we *had* been "living much luxury" at Benteng Tinggi.

Take seven very individual, independent women and one gentle-man, accustomed to being a leader, put them in cramped quarters such as these in which we were now being confined, and what do you have? Put God in the midst, and you have that rare and beautiful thing known as the fellowship of the saints.

We set up a work roster. Firewood had to be gleaned from the jungle, water carried from a nearby stream for cooking and washing clothes, edibles collected and cooked. Each person was responsible for his or her own bed. A place was found for everything, and everything had to be kept in its place. By turns we swept and cleaned the floors and bathroom.

In Malino we found a type of wild millet that we dubbed "bird seed," because we always sang as we cooked it. Whatever it was, we were pleased with our find, for the more water we added and the longer we cooked it, the more it swelled. Our stomachs were full. On many occasions I stood at the fire, stirring by turns with Ruth a large pot of this bird-seed porridge. We would speculate on where Ernie and Russell were in Pare Pare, what the conditions and food were like, and run through our repertoire of love songs until the porridge was ready.

Having no kerosene left for the lamps and no candles, we made torches from the silk of the milkweed pod and crushed kemiri nuts. This type of torch smoked badly, but it furnished light by which to work and read.

While gathering wood one day, Miss Seely ran a very large sliver into the top of her foot. She had difficulty removing the piece of wood, and eventually the foot became angrily inflamed and swollen to three times its normal size. We feared gangrene and watched an awful red streak appear beneath the skin. That signaled danger, but Miss Seely exhibited perfect calm. After soaking her foot, she covered it with a clean handkerchief to keep off the flies. To our inquiries she replied, "I'm trusting the Lord. He will heal it." Miss Seely uttered her daily proclamation with total confidence and, just as she had believed, within a period of time her foot returned to normal size.

We were not left to our own devices many days before soldiers came to check on us. They gave the house a thorough inspection, shaking, lifting, and feeling every article in the house. When the inspection of Dr. Jaffray's bed started, I turned away lest my face betray my agitation. No one moved until the clump-clump of their heavy boots could no longer be heard.

"Your flashlight, Dr. Jaffray?"

"It's here. They lifted my mat and shook out my blanket, but never touched my pillow." Neither the flashlight nor the watch had been disturbed. I found joy and quietness of spirit in letting God be God. I never needed again to ask if the flashlight and watch were there.

During this time we wondered why the Japanese had moved us from Benteng Tinggi. We were closer to Malino, but still too far away for them to keep a close watch on us. Perhaps the army intended to use Benteng Tinggi for a rest and recreation center or for some military purpose.

We remembered seeing some of these houses through binoculars from Benteng Tinggi—"So on such a clear afternoon, Benteng Tinggi should be visible from the end of the peninsula just beyond the last houses," said Margaret Jaffray.

Dr. Jaffray walked a short distance with Margaret and me, then turned back. We had not been forbidden to walk wherever we wished but had stayed close to the house purposely, not wanting another encounter with the Boegis people.

By the time we reached the last of the houses, some of the buildings at Benteng Tinggi were visible. There was activity about the property, so we knew it was being utilized by the Japanese. That was admittedly better than having the conference center vandalized by local Boegis marauders.

Knowing it prudent not to be gone too long, we turned to go and came face to face with a Chinese, who was as startled to see us as we were to see him. He greeted us cordially, then disappeared down the jungle path from which he had just emerged. Some instinct within me whispered, "You'll hear from that man again."

CHAPTER FIVE

In May of 1943, five months after our deportation from Benteng Tinggi, Miss Seely saw one of our Macassar friends while collecting rations at the native market in Malino. He relayed the news that Ernie and Russell were no longer in Macassar; this we already knew. No one had been able to learn anything further about our husbands since the civilian men's camp had been relocated in Pare Pare.

Standing in the lean-to cooking and stoking the fire one day, I picked up a familiar clump-clump-clump of soldiers' feet. Hastening to the nearest window, I called in, "Soldiers! They're headed this way!" With a great pretense of surprise at seeing them, I obeyed when a soldier motioned me into the house. The one whose Indonesian was the least bad gave us instructions to pack. We were restricted in how much we could take; there would be no carriers this time. The soldiers continued on to the other houses to inform our neighbors to pack. We ran to the cooking lean-to, doused the fire, then hastened inside to pack. Wearing all my lingerie and four dresses, one over the other, I made room for other things in the briefcase.

Arriving in Malino just before sunset, we were ordered to go to the church, where we would spend the night. The church was already filled to capacity with people and their luggage. We found places for Dr. and Mrs. Jaffray on a pew. It was a hard bed, but at least they could stretch out after the long, exhausting march of the afternoon. The rest of us tried to make ourselves comfortable on a pew or the floor in as close proximity as was possible, lest we become separated when the soldiers started loading the trucks. All that long night we could hear the rumble of arriving enemy trucks and the screech of brakes as they stopped alongside the church.

There was little sleep for any of us. Little ones clung to their mothers and bewailed the unnatural, uncomfortable atmosphere. Fear is infectious—now I understood why the Lord had instructed the army leaders in Deuteronomy to send home the soldiers who were fearful, lest they cause the others' hearts to faint as well. The children, sensing the fear of their mothers, cried for what they did not understand, while their mothers cried for what they *did* understand—that all our lives hung by a very slender thread. We were at the mercy of capricious, unprincipled men.

Some of the women were leaving behind graves. They clutched their small bundles to them and rocked back and forth, cheeks wet with tears, eyes closed, voices mute from sorrow too fresh to be expressed audibly.

"My Lord, do You not hear the cries of the little ones? Who will be their defense? Rise up and avenge us of our enemy." It was a night in mournful Kadesh, grieving for what had been, was no more, would never be again. Weeping endured the long night through, followed by a morning from which joy had fled. My head pillowed on the briefcase, I studied the faces of those around me. Fingers of first light had reached through the windows and softly within my heart I began to sing:

> In the glow of early morning
> In the solemn hush of night;
> Down from heaven's open portals,
> Steals a messenger of light,
> Whisp'ring sweetly to my spirit,
> While the hosts of heaven sing:
> This the wondrous thrilling story:
> Christ is coming—Christ my King.
>
> Oft me-thinks I hear His footsteps,
> Stealing down the paths of time;
> And the future dark with shadows
> Brightens with this hope sublime.
> Sound the soul-inspiring anthem;
> Angel hosts, your harps attune;
> Earth's long night is almost over,
> Christ is coming—coming soon.

"Yes, come, Lord Jesus, come quickly." And I listened for His footsteps, for never during our imprisonment thus far had the future been darker or the shadows deeper. It had now been more than a year since Russell and I were separated, more than six months since there had been any definite news concerning him. Every move meant greater separation. Now where? Unless we were to be taken to Macassar, where besides Malino could they find housing for more than a thousand women and children? I touched Ruth and suggested that some of us should go to the makeshift washrooms while others watched the luggage. Dr. Jaffray, the early riser, was already awake. We checked and arranged our baggage so that nothing should be left behind, then distributed Dr. and Mrs. Jaffray's personal belongings among us. Breakfast was a meager portion of bread.

A truck roared to life, frightening the children into wakefulness. People began to mill about; little ones strayed from their mothers. When the moment of realization came that they were lost, they began to scream, *Moeder! Moeder!* "Mother!" It was pandemonium. Other drivers started to warm up the engines of their trucks, drowning out the sounds of human misery. The loading began without consideration for the elderly, the infirm, or mothers with children. Those who had been in Malino were apprehensive lest one of the officers be the former owner of Kaneko's Variety Store. He had reappeared as a captain in the army. He made a great production of beating all the women he felt had been rude to him while he was masquerading as a shopkeeper. Everyone walked softly before him in his true role as army officer.

A fixed bayonet and prior knowledge of our captors were great persuaders. The trucks were loaded. Much prayer was made that no one would be thrown out of the stakebed truck. It had no tailgate. Those of us who could secure a firm hold on a support rail allowed those in the center of the truck bed to hang on to us. We held our bags between our feet. The drivers whipped around bends with no regard for their human cargo. They slammed on their brakes to throw us against one another. The abrupt starts were the worst, however. We dreaded that a child or elderly woman would be wantonly tossed off the end.

The sun had no pity on us either. The heat became oppressive as we reached sea level, and we were all candidates for sunstroke. I felt bruised, battered, and at the boiling point—I and my four dresses and layers of lingerie.

Things began to look familiar. We turned off the main road and drove along between *sawah*s, "flooded rice fields." To the right we saw a large moat with a barbed-wire fence on the far side.

"So this is it—Kampili, the native tuberculosis sanitarium." Everyone nodded, too exhausted to speak. The driver turned to cross the moat and bring the truck to an abrupt stop inside the barbed-wire enclosure. Those near the back jumped down to help others. I experienced the sentiment expressed in the spiritual: "Sometimes I feel like a motherless child—a long way from home!" Iowa was never like this—the area, denuded of vegetation, was a plain of red clay cluttered with partially completed structures resembling Dyak long-houses, except that they weren't standing on piles. To the right were the cement houses of the sanitarium.

We were herded toward the farthest block of six barracks. The first four buildings were complete. Barracks 5 and 6 were nearing completion. A young Chinese was inspecting the work of his crew of carpenters. I saw him speak to a Dutch woman, who turned and pointed directly at me. I nearly swallowed my tonsils from shock. I had never seen the man before. Steadily moving in my direction, he gave suggestions here and there to the workmen. One complete turn, presumably to inspect the building, told him that there were no Japanese in sight, so he boldly handed me a letter, saying it was from my friend Wiesje, and did we need anything?

I crumpled the letter into my hand and, turning to look at Margaret Jaffray so that the Chinese and I would not appear to be talking together, I thanked him and Wiesje but urged please, not to risk another letter. It was far too dangerous for them both.

"Come, Margaret, let's find someplace where I can get out of these clothes or I'm going to die of heat exhaustion. Did you ever see a place that reminded you more of the 'abomination of desolation' than this?" We found a block of toilets at the back of the barracks. Margaret stood guard, as there were no doors and workmen were everywhere. One, two, three, four—off came the dresses, then the redundant lingerie. It was such a welcome relief.

Margaret and I then read Wiesje's expression of the family's concern. She wrote that if we were in need of anything we should tell Suii; he was a good friend and we could trust him. This I appreciated, but we hadn't the slightest intention of endangering their lives. With the letter stashed inside my clothes, we joined the rest of our group. I would need to find a fire where I could dispose of the letter after the others had read it.

We learned that people from Ambon and nearby islands had been brought to Kampili, after their camp in Ambon had been bombed by the Allies. Many had died or been injured as a result of the bombing. The survivors of the bombing were all billeted in the cement houses that had been part of the former tuberculosis sanitarium. Dr. Jaffray was directed to one of those cement houses, where he would be lodged with the other men and all boys over sixteen.

As soon as the trucks carrying the construction men left for Macassar, we were told to choose a place to sleep in Barracks 6. We first examined our barracks. The walls of these half-block-long structures were woven bamboo mats, their steep hip roofs were covered with grass, and the floors were bare dirt. Extending down both sides were double-decker bamboo sleeping racks. There was a space between the top of the wall and the roof, providing light by day and cross-draft at all times. How much human tragedy or triumph these barracks at Kampili would witness was a question that time alone could answer. With the prospect of moving ahead of us, we made no attempt to settle in.

The sanitarium kitchen, which we experienced for the first time that night, had been enlarged to facilitate cooking for the whole camp. We had some skilled cooks who knew how to artfully use spices and make the unpalatable palatable.

With the tying of the last piece of rattan and the completion of all twelve barracks, orders came to move all who were not Dutch to Barracks 8. Gathering for the first time, we found that we were a conglomerate of many nationalities. We dubbed Barracks 8 the Heinz Barracks, for we were nearly fifty-seven varieties.

Each barracks chose its own leader to represent them to Mrs. Joustra, the Dutch head of the camp who was our liaison with the camp commander. Unless languages were a factor, I have no idea why I was ever chosen for barracks leader. All dialogue with the Japanese was in Indonesian, and with the women and children of the barracks, English, Dutch, and Indonesian were used. In all of these, I could express myself with a measure of fluency. Yet even in the English-speaking block, we represented Canada, England,

Scotland, Ireland, Germany, Armenia, Israel, Czechoslovakia, and Russia. In the Dutch-speaking block, we had a number of Eurasians and a German. Three Chinese families, a Menadonese, a Macassarese married to a Dutchman, and an Orthodox Jewish family communicated in Indonesian, because no one else knew their mother tongue.

The camp commander made a nightly inspection tour, beginning with the ringing of the bell at seven o'clock. The barracks leader stood at the door and called the orders: "Attention," "Greet," "At ease." When we bowed, we greeted the commander in unison with "Good evening, sir." This he acknowledged with a salute. We all tried to be present, for we never knew when he would turn in and check each bed. The punishment always exceeded the crime.

Immediately following evening roll call, everyone came to the front section of the barracks to hear the announcements and check the work schedule for the following day. The first night in our barracks, I established a practice that I believe was responsible for maintaining the high level of compassion and cooperation that existed in our small community. That night and every night, we invited everyone to remain while we read a portion of God's Word and prayed. We were united by a recognition of a mutual need from within for help from One Who is greater than we. We faced a common enemy from without, and if we were to survive we had to function as a unit. The interpersonal barriers of language, race, and color became nonexistent, and an ever-increasing appreciation of one another enabled us to face with courage the common plights of most prisoners of war: suffering, hunger, deprivations of every kind, forced labor, bombings, disease, psychological pressures, death, and lonely graves. People from other barracks often joined us during evening devotions. Throughout those very difficult years that tried our souls, God kept our barracks a calm center in the eye of the military storm that raged around us. There was a sharing, a concern, and a love that was unique. We struggled to preserve family feelings, to discover ways to lift morale, to encourage, comfort, and bear one another's insupportable burdens. I am convinced the harmony we experienced in Barracks 8 was due to the spiritual shelter beneath which we all hid when there was no other refuge.

As soon as beds came free, Dutch families asked for transfer to Barracks 8. They were people with a heart for God, and they were much appreciated.

When the morning bell was sounded at seven, it was the signal to begin the day's work. Each barracks had a work quota. A certain number of women had to be assigned to working in the central kitchen, hospital kitchen, and camp gardens; and to clearing land, felling trees, working on the roads, unloading trucks, raising pigs and chickens, pumping water, sewing, knitting, cooking porridge, boiling water, doing hospital duties, and nursing. The sweat of our bodies and the craft of our hands were required to keep ourselves alive and to advance the Japanese military machine.

In order to give variety and change, I arranged the roster for the week so that each woman performed a different type of work each day. In this way each one came to know what was involved in the various types of work. We soon settled into a pattern. Recognizing that the physically weak ones could never have performed the heavy work, the younger and stronger women accepted the responsibility of work at the piggery, in the garden, and on the road, as well as the coolie work. It was immaterial who did the work, as long as our production didn't fall behind.

In the early weeks, the morning porridge was cooked in the large central kitchen. Cooks were responsible for fetching the rice from a large supply *godown* that had been built in the open space in front of Barracks 7 through 12. The rice had to be cleaned and washed several times before it was set to soak in the lower half of a large fifty-five-gallon drum. By 4:00 A.M. the cooks had to be on duty, stoking their fires and, with large wooden paddles, stirring constantly their large containers of gruel. It was time-consuming, but so was waiting in the long line to get our dipper of rice porridge.

Porridge from one drum was especially good one morning. In fact, some said that it tasted like chicken and rice soup. As the day grew lighter, it was discovered that the delicious brew was bird, rat, and rice soup. Knowing that birds roosted in the window-like openings that were smoke escapes, and that rats were multiplying and on the prowl, we deduced that a rat had attacked a bird and in the struggle both had fallen in the drum and drowned. If the rat's tail and hairs and bird feathers hadn't eventually been dished up, no one would have known—or lost her breakfast.

A long shed containing eating and open-fire cooking areas had been erected in front of each block of six barracks. Realizing how inefficient the breakfast arrangement was, the camp commander issued a half kerosene drum to each barracks, and one of our young women with a small baby volunteered to be responsible for the porridge. Our drum contained the taste of kerosene even though it had been scoured with wood ash many, many times. It was a long time before our porridge was completely free of the acrid, undesirable flavor.

There was some changing of beds in the beginning day or two, with families and close friends choosing beds in close proximity to one another. On the righthand side by the door, Lilian Marsh had claimed the top rack. Miss Marsh's presence had a stabilizing influence. Her weight had dropped dramatically, and a hacking cough tormented her; yet despite the fact that she was physically weakened, she never complained or sought privileges because of her condition. Her hands looked like bird claws. We were greatly concerned about her condition. I was able to arrange a job for her in the hospital kitchen, where she could sit most of the time while working. One day when she came home, grubby from washing pots and pans, I called her Cinderella, a pet name that stuck.

In the rack beneath Lilian slept Mrs. Woodward, also English, who had worked with her husband in central Celebes for the Salvation Army. She was a lively, bright presence, a very motherly type, though she had no children of her own. A lot of her time was spent assisting Mrs. Snaith, the neighbor in the next tier of racks, to raise her daughter.

Mrs. Woodward and Mrs. Snaith had much in common—both were associated with the Salvation Army, both had the misfortune of having ill-fitting dentures, both worked in the diet kitchen, and both were godly women who loved the Lord. Mrs. Snaith was Irish, and her magnificent brogue enhanced many a conversation. Her daughter, Emerald, was a slender, tall child with straight brown hair; she moved with the grace of a gazelle. In spite of her youth, she was never rude or petulant.

Bright spots of fever always showed in Mrs. Snaith's cheeks; she suffered from weakened lungs and ulcerous sores on her feet and legs that refused to heal. There was little we could do, lacking sulfa powder and the proper nutrients. Our only recourse was to cover the ulcers with an application of a strong soap solution, then

with clean rags now turned gray but sun-sterilized. The rags pre-
vented the flies from spreading further infection.

One day Mrs. Snaith and I were on the "honey dipper crew,"
emptying cesspools, and she slipped into that thick, putrefying
green filth, which flowed into her open ulcers. This would have
been a trying experience under normal conditions, but this woman
had enough to worry about healthwise without the possibility of
further infection. Big tears rolled down her cheeks. I walked her
to the well between the barracks and drew many buckets of water
to wash the filth from her legs and cleanse the sores. Then, feeling
that she really had had enough, I put a strong soap solution on
her legs, bandaged them, and sent her to bed.

Next to Mrs. Snaith and Emerald lived Oma Zimmerman. *Oma*
is the Dutch word for "grandmother." Mrs. Zimmerman was a
widow whose husband had been a German planter, and she re-
galed us many a lonely night with tales of growing up on a
plantation in Java. Although she was now about seventy and tooth-
less, traces of great beauty hung about her. Her long brown hair,
beginning to be frosted with silver, fell in waves, and she wore
gold-rimmed glasses perched about halfway down her nose (as
she could see better at a distance without them). With her high
Eurasian cheekbone structure and great vivacity, it was easy to see
that she had been very, very beautiful.

Oma Zimmerman was a willing helper. She was also a natural-
born cook—an artist in the kitchen under the best of circumstances, an ingenious inventor under these, the worst. She had the
ability to make the few things she acquired taste borderline gour-
met by adding whatever herbs or condiments she could scrounge.
We combined what coffee grounds we had amongst us and Oma
would cook these, then carefully spread them out in the sun to
dry. I can't possibly guess how many times this process was repeated, but with her gift of taking nothing and making it taste
better, we enjoyed coffee for quite a while. We ate rice at noon
and evenings, but Oma always managed to scrounge something to
go with it.

Oma's daughter, Maus Dol, was very tall—about five feet, ten
inches—and in her early twenties. She was large, Junoesque, and
very beautiful, probably echoing her mother's youthful image. She
had real physical strength, a rare commodity in Kampili, and often

assisted me in extra tasks with the same positive spirit that exemplified her mother. Maus's little daughter Paxje ("little Pax"), lived in this tier with the grandmother and mother. A little towhead with pigtails, she must have resembled her father. Paxje sucked her index and middle fingers so much that her front teeth were slightly protruding.

Next to Maus, Pax, and Oma Zimmerman lived Maus's friend, Mrs. Verschoor, wife of a patrol officer, a young, petite Dutch lady with a little girl, Anneke, about Paxje's age. Anneke and Pax had both been accustomed to the role of "top dog," so by turns they fought like tigers, played together like angels, and were inseparable. Mrs. Verschoor and three other Dutch families came to our barracks by request. They were all real assets, and we had cause to be grateful that they joined our "family."

Across the aisle lived another Dutch family. Mrs. Routs was a fine Christian woman, and I imagine she felt a kinship to the believers who lived in Barracks 8. She was very tall and perhaps before the war had been overweight, but of course those hard days had melted away the excess. Her work responsibility was on the garden crew. She had two dark-haired boys who were very lean, very typically Dutch. The littlest one, about two years of age, was rarely without the index and middle fingers of his left hand in his mouth. No one discouraged or made fun of finger- or thumb-sucking, however. The children needed all the comfort they could get. He was called Broertje, which means "little brother." Older brother adored him and cared well for him.

Beyond the Routs family lived the van der Kleys. The mother was also a Christian woman with two children, Ansje and Susje. She was refreshingly frank but didn't begrudge the other person his opinion. She and Mrs. Routs had long been friends. She, too, worked on the garden crew and was always available to fill in for a sick person.

Lieke Marcar was an attractive, amiable Dutch woman much concerned with the welfare of her family. Her son, Dougie, and her mother-in-law, Oma Marcar, were with her in Kampili; her husband had been in the import-export business in Macassar and was now interned with the other civilian men in Pare Pare. Dougie, like the other young men of our barracks, made a great effort to adjust to our matriarchal environment, and I'm sure his father was proud of him. Oma Marcar, another of our grandmothers, was quite tall, beginning to stoop, her step slowing and her hair quite gray; but from her present bearing one could imagine that in her younger years she must have been a very regal, stately person. Oma's needles moved unbelievably fast as she knit white three-quarter-length socks for Japanese officers. Oma was Armenian, and her dresses were longer than usual. I compared her in my mind with Whistler's mother; all she needed were the rocking chair and the white cap.

Next to the Marcars were Tiele Noll and her small son. Tiele was our capable porridge cook, up every morning at four. Tiele's hair was long, black, and wavy. She had a sister who lived in Barracks 6; Ellie was very like their Dutch mother, while Tiele favored her Armenian father. Tiele's eighteen-month-old son was a beautiful child; consequently we all petted and spoiled him properly. Tiele was a good mother and requested the middle-of-the-night work so that she could be with Wiwi (pronounced Veevee) during his waking hours. When the toddler developed a large abscess on the forehead just below the hairline, it became necessary to have it lanced. Tiele gave him to me and walked away: "I couldn't stand to see it done. Please hold him for me." There was no local anesthetic to dull the pain, and it fell to my lot to hold Wiwi with his arms pinioned to his sides while the doctor opened the inflamed swelling with her very dull scalpel. Wiwi's look of shock turned to one of intense pain, and big tears rolled down over his little cheeks. As the doctor mopped up the blood and pus, then bandaged the wound, I held Wiwi close and tried to comfort him. A great lump of hatred for the indifference of my captors to the sufferings of the innocent little people nearly choked me. "It's all right, Wiwi. Let's go home to Mother." I was thankful that the sobs had subsided before I delivered him to Tiele.

Across the aisle from Tiele and Wiwi was an Armenian family; the father and husband was interned at Pare Pare with Ernie and Russell. Carr M. David was a gentle man of Iranian descent who had been the Honorable British Vice-Consul in Macassar before the war. His good breeding was reflected in his family—a daughter, Elsie, and two sons, Ronnie and Arnold. Rose, the wife and mother, was a woman of dignity. Rose's hair, now graying, had been jet black like her daughter's. She wore it parted in the middle. She never needed rollers or pin curls, for her hair had narrow, natural waves. It was beautiful, as was her skin. We all appreciated Rose; she was my capable assistant, whose advice I valued, and our friendship has been a lasting one. Elsie was a happy-go-lucky teenager, always able to respond with a smile, a joke, or a laugh to cheer others. The patients at the hospital, where she worked, needed someone of her temperament. The boys were opposites: Ronnie was tall, light-complexioned, and outgoing, while Arnold was short, dark, and shy.

Saartje Seth-Paul and family occupied the next two sections of beds. She was a good-natured, outgoing person who never allowed anything to ruffle her calm exterior. Her sister Bea de Graaf was very like her. Both were strong; both were always helpful. Bea worked at the piggery and Saartje was a member of the garden crew, as was Jet Robyn, her other sister.

Oma de Graaf, mother of Saartje, Bea, and Jet, was Menadonese and continued to dress in the traditional, very comfortable sarong and overblouse. Like Oma Zimmerman, she wore round, gold-rimmed glasses that were usually midway down on her nose. Even in a prison camp Oma was lovely, growing old surrounded by her daughters and their families.

Two of Saartje's three children were twins. Paula was beautiful and gave promise of becoming quite tall; already she was taller than her handsome twin, Pim. They were lovely complements to each other, and when one contracted a very serious case of hepatitis, so did the other.

The youngest Seth-Paul child was Mary, who could be a little terror, because she was so adorable that everyone gave her whatever she wanted, if it were to be had.

Behind this family lived a native woman from Macassar, Mrs. Weidema. She had a boy about seven years of age, Neilus, who was difficult to discipline. She also had a baby, born shortly after our arrival in Kampili. When the camp was struck by a gastroenteritis epidemic, Baby Weidema, too, became extremely ill. One evening during our prayer and Bible study, Mrs. Weidema came running from the hospital. "*Njonja,* please come and pray for my baby," she pleaded. She had been going regularly to nurse the baby, but her milk had dried up and the baby was not responding to the rice water they were using as a substitute. I went back with her,

my heart torn by that limp and lifeless bit of humanity. We bowed
our heads together, and I prayed that the Lord would spare her
baby.

Returning to the barracks I was met by another young mother
from our barracks, Lennie Ripassa. She had an infant and was
nursing. "*Mevrouw* Deibler, I have more milk than I need," she
volunteered. "I would be willing to breast-feed the other baby,
too." Which is exactly what she did, feeding the sick, starving baby
until he regained strength and both babies were weaned. This was
God's answer to prayer. Both of the newborns from our barracks
survived.

The mother of the last family on the right side of the barracks
was a woman of many faces. Often in the middle of the night, we
would hear the children crying and crying. Finally I would drag
myself from the rack and go down to find the three little tykes left
alone, their mother nowhere to be seen. One would have awakened
and found the mother gone, and his or her crying would have
awakened the other two, who joined in. Not once but many nights
this happened. Once when I checked on the children and discovered that they were covered with their own filth—they had defe-
cated in their beds and rolled in it—I was furious. I suspected
the woman was with one of the Japanese officers. Her stock excuse,
that she was sick and had been in the WC, stank.

That night I waited for her. By the time she sneaked through
the door, I had a full head of steam on. It was bad enough to miss
my own sleep with the workload of tomorrow facing me, but I was
most incensed at the neglect of those little children. I addressed
her in no uncertain terms and informed her that I would have to
report this infraction of the rules to Mr. Yamaji. She was as con-
genial as if I had been telling her what a lovely night it was! Her
response was, "Yes, I can see you have to do that."

I was totally deflated. How do you fight someone who lies
down and says, "You win; I'm dead"? That was one face I hadn't
seen before. I had never asked her help but what she hadn't gone
the second mile: "Is there anything else I can do to help you?"
Another face. When she was there, she was a good mother; the
little ones loved her.

The next night she went off again. This time I felt something
had to be done, for the sake of the children and of others whose

sleep was disturbed. I was sure she wouldn't be hurt, but I wanted to register my disgust and protest. I hoped the camp commander would be ashamed and not see her again. At roll call I told him I wished to see him when he was finished. We walked to the back of the barracks together.

The whole affair was a slapstick comedy. "This woman has not adhered to camp rules to be in quarters after lights out. Her children are abandoned during the night, leaving the responsibility to others in the barracks. This has occurred numerous times." Yamaji pretended to speak harshly to her, swatted her across the backside with his cane, and ordered her to stay in the barracks. She said she would and grinned at him, because she knew that I knew where she went at night—to his quarters.

When this woman was taken to the hospital ill with hepatitis, Ruth Presswood, the compassionate nurse, went into the last tier of racks and cleaned out the whole area, digging out the dirt floor where the children had urinated and done their bowel movements. The smell alone was enough to knock you over. She washed everything down and bathed the children. It was a mystery to the rest of us how in the world such winsome, good, tractable children could have belonged to that woman, and how she could have endangered their future happiness if she really cared for them. As I said, she was a woman of many faces.

I felt great pity, not only for the children but also for the mother. Someday, God willing, the war would end. What then?

Perhaps the fact that there were those who would sell themselves for special favors spared the rest of the women from being molested; some were quite beautiful. It was hard to understand the psychology that motivated a woman to willingly become the camp pariah—it was a standard of value that overwhelmed the rest of us. I learned why the woman being taken to Malino had protested so vociferously while we were still imprisoned at Benteng Tinggi. She had been living with the Japanese and hadn't wanted to face the other Dutch women.

On the left side of the barracks lived a Jewish family, the mother's name was Rachel. She had five children: two girls and three boys. Mr. Meshmoor had been a merchant in Macassar. Four children were in the camp with us, but the oldest son, Raimond, had been taken to Pare Pare with the men. She was an orthodox

Jewess, and we were able to arrange permission for her to cook their portions of rice and vegetables in her own pots and pans so that they would be kosher.

Rachel was a sweet woman. Because she suffered severely from asthma, it was necessary to exempt her from all work. One night the older daughter, Esther, came running. "Come quickly!" she pleaded. "Please come look at Mother." I ran down to the end of the barracks and saw that Rachel was gasping and fighting for breath. I immediately sent for others to help me carry her to the hospital, where she was given shots of some sort—we couldn't trust the supplies being brought in. She didn't respond, and I was sent for again.

I ran into that long ward and knelt by her bed as she struggled against suffocation. "Rachel," I whispered, "we pray to the same God. I'm going to pray for you now." She nodded her head in agreement. I laid hands on her and prayed in the name of the God of Israel and His Son, Jesus Christ, that the Lord would touch her body. Immediately her breathing eased, and the following morning she returned to the barracks.

On occasion the women were allowed to write letters to their husbands. Because Rachel didn't know how to write, she dictated and I acted as her scribe. Her letters conveyed compassionate love for her husband and son. (Her concern for the children still with her was obvious.) She always ended these missives with "1,000 kisses and hugs." Often she would share with me a portion of the rice that was left from the bottom of the pot, crispy and browned; it was a treat. Whenever we had finished speaking, Rachel kissed me on both cheeks. A tremendous bond developed between us, enhanced by our mutual devotion to God.

Advancing along the left side of the barracks, we came to the beds occupied by Mrs. Muller and two little ones. Mrs. Muller was very, very small, Eurasian by birth, schooled in Dutch, and she was an altogether dear person. She looked as if a breath of wind could blow her away, and when she volunteered for work in the garden crew, I told her so. She laughed and said, "Oh, but I'm very well, *Mevrouw.*" She was strong and cared well for her two children, who looked like tiny dolls. She must have been very happy in her marriage, and I never ceased to appreciate her optimism and cheerful countenance.

Next to her were the three Chinese families from Macassar. All of them had been connected with the Chinese embassy. Mrs. Wang Teh Fung was older and worked in the hospital kitchen. Her fine sixteen-year-old son lived in the house with Dr. Jaffray. Both Mrs. Lie Tsu Hawi and Mrs. You Pao Heng asked to work with the garden crew. It was an interesting picture to see Mrs. Lie and Mrs. You, large hoes over their shoulders, leaving for the fields each morning—followed by their assorted cluster of five little girls, all wearing sun hats like their mothers.' The Lies had been vibrant members of our Chinese church in Macassar.

Next came three Armenian families. The husbands of all three households had been in the import-export business in Macassar and were now incarcerated in the civilian men's camp in Pare Pare. Serah Paul with her son, Freddie, and daughter, Dolly, had the block of beds next to Mrs. Wang.

While in Malino, Dolly became very ill, and the doctor was sent for. His tests proved that her illness was indeed meningitis, as he had suspected. She became paralyzed from the waist down and, even though she recovered from the infection, she went into deep depression. It was not difficult to understand how traumatic it must have been for a pretty young girl like Dolly to find herself paralyzed and see her legs not developing as they should. I found instant rapport with Dolly and tried to see her often. A woman in the next barracks had also had meningitis, however, she had learned to walk again with the use of canes. Because their experiences had been similar, she took a vital interest in Dolly, suggested exercises, and encouraged her to believe that she, too, would recover the use of her legs. The two of them worked together as much as possible. With everyone in the barracks lending encouragement, Dolly progressed to the point where she could walk holding onto the ladders of the top sleeping racks. Swinging first one leg then the other, Dolly moved up and down the length of the barracks. With this remarkable improvement, her depression lifted. A wounded spirit and a broken body began to mend.

Freddie, Dolly's junior by two years, was very good for her. He never treated her like an invalid. Their occasional disagreements were good therapy. Freddie was tall like their mother, Serah. When we needed another woman for the garden crew, Serah volunteered. She had suffered much during her daughter's illness, which would account for her pensive, often sad expression. She was not moody, for the moment she was spoken to, her face lit up and she responded. She must have been comforted often by Freddie in her concern for Dolly. There was a strong bond between them.

Another family, also named Paul, occupied the next two beds: Lena and her daughter, Vonnie. Lena's son, Robby, was housed with the men. For whatever reason, the mother-daughter relationship seemed to have suffered. They needed each other very greatly, but somehow they appeared to have lost the ability to understand and communicate with one another. It was one of the unfortunate consequences of the splitting up of families during the war, made the more regrettable because both Lena and Vonnie were basically nice people. Initially Lena had been a large woman, but with the passage of time there was less and less of her. She, like Serah, worked with the garden crew.

Next came Lucy Galstein, daughter Joyce, and son Herby. Every one of them had lovely, wavy, brown hair. The beautiful hair seemed to be characteristic of the Armenians. Lucy was tall and Joyce was going to be, while Herby was short, a fact that troubled him not a little. Lucy's mother, Oma Galstaun, had the bed next to her. Oma knit white socks for the officers, and Lucy belonged to the garden crew. There were frequent disputations between the two women revolving around how to raise Normie Galstaun, the fifth member of the family. He was Oma's grandson and Lucy's nephew. Whatever their recipe, I think the ingredients were correct, for with a full measure of Oma and several dashes of Lucy, Normie turned out to be a tall, garrulous, delightful, slightly bowlegged imp. Normie and Herby kept things interesting. When Normie ran out of ideas, Herby pulled a few out of his bag of tricks.

Then we had another fine Eurasian family of five by the name of de Keizer. There were two boys and two girls. Jan was the oldest; he was tall and quite thin. Carolina was fairly stocky, built very much like her mother. Both she and the younger sister had long brown hair that was always neatly braided and tied with ribbons. Carolina, who was about seventeen, chose to work at the piggery with others of her age group. Bertie, the younger son, always had a grin and a good word for everyone. He was cooperative and helpful to his family. Mrs. de Keizer was on the staff at the central kitchen. She was quiet, very proud of her family, and an altogether fine person.

The next section of racks was occupied by Oma de Vries, a Eurasian woman married to a Dutchman. Her lovely daughter, Lennie Ripassa, was about eighteen years old. Lennie was married, with a little daughter, Iris; her second child, Joyce, arrived soon after Lennie came to Kampili. It was this young woman who volunteered to nurse the other infant in our barracks. We tried as much as possible to provide her with extra food, because of the additional responsibility she had undertaken. Oma suffered from diabetes and appeared to be morose, probably due to her physical unwellness. She was a Christian woman and very faithful in her attendance at devotions.

Mrs. Trauerbach had the fourth bed in this section. She was a devout Christian, helped Oma care for Iris, and, like Oma, worked in the hospital kitchen. On down the row was Ellie Kucerova, a Czechoslovakian girl with an effervescent personality, complemented by a beautiful singing voice. She was a cheerful, positive person and much appreciated by the patients in the hospital where she worked. As I came to know Ellie better, I realized that beneath this happy exterior was a person with a very deep hurt. Ellie was adept at appearing carefree and lively, but beneath the surface I had seen a very grieved spirit. She was a devout Roman Catholic, and we had many opportunities to talk together about God. Ellie shared with me the source of the hurt and allowed me to pray for her.

Beneath Ellie was a German girl named Katie Eckstein, for whom I felt an immediate affinity. We were about the same age. She had come from Germany to live with her sister, who was married to a Dutch patrol officer. Katie and Ellie became close friends. I felt joy to be available whenever either of them needed to unburden themselves to someone. In a measure, God enabled me to bring help and comfort to them on a spiritual level.

Not only was Katie physically beautiful, but she had inward charm and grace. She was always a willing helper and a clever, creative seamstress; so naturally she was assigned to the sewing room, making uniforms for the Japanese and work clothes for us women.

The section next to Katie and Ellie was occupied by Mrs. van der Haarst and her four daughters. Shortly before the Japanese invasion, the van der Haarsts' oldest daughter, Leintje, suffered a mental collapse. The doctors could find no reason for this sudden personality change; seemingly overnight, a beautiful, cheerful, out-going girl became depressed and withdrawn, except for occasional unpredictable fits of temper. Leintje had grown obese and had become exceedingly strong. The whole family was under a pall of disbelief that this had happened to their beloved Leintje.

The mother showed me a picture of Leintje before the collapse. She had truly been a lithe young woman, with a flawless complex-ion, natural light blond hair and dark, dark blue eyes. In a few months time, she had totally regressed. Though quiet most of the time, when she was enraged only Mrs. van der Haarst could handle her. One night a terrifying scream rent the barracks, and we ran to where the van der Haarsts lived to find Leintje trying to choke her mother. It took her sisters and a number of us to force them apart. More than once, she made attempts on Mrs. van der Haarst's life, once trying to stab her with a knife. How often I saw the mother standing with the picture, grieving for her Leintje, and how often I heard her say, "If it wasn't for God, I don't know what I would do." She loved the Lord and drew upon Him for help and comfort—what a great wound this family had sustained, a hurt only God could heal.

The next daughter, Tina, was a physically strong, very dear girl with red hair. She worked at the piggery. Mien, the third daughter, was very tall, and very lovely, with blonde hair. She worked with the garden crew. The youngest daughter, Katrina, was spoiled, naturally, but she was only thirteen. They were girls to be proud of, and they brought much comfort to their mother. This family was respected by all of us, and we sorrowed for them and Leintje.

During a severe outbreak of dysentery, Leintje took ill and died in the camp hospital. The family could but say, "It was a blessing," since the doctors held out no hope of Leintje recovering from her mental illness.

Next to the van der Haarsts came the section of the barracks where Miss Seely, Mrs. Presswood, Miss Kemp, and I lived. Mrs. Jaffray and Margaret lived in Barracks 6, where Mrs. Jaffray had many friends from Macassar; however, they later moved to one of the cement houses, which was better for Mrs. Jaffray as it was less damp and not as cold at night. Basically, this was the arrangement that existed at the end of the war.

A very large *godown* had been constructed in front of our block of barracks. Here were stored all the camp supplies. The rainy season came with a promise that we would be sloshing around in a sea of red mud until the dry season came again. The grass roof became waterlogged; the bamboo frame was not sufficiently braced to withstand the wind and driving rain. One afternoon we heard a crackling sound, followed by a mighty swoosh— then *crash!* A cry went up on all sides for everyone to get out and help salvage the food. The whole camp worked for hours, trying to save as much as possible. Ruth and I were given a rigid bamboo stretcher on which to carry between us heavy copra bags full of saturated rice and sugar. I called on the Lord for strength, for I was sure my arms must come out of their sockets. My spine felt as though the vertebrae were being crushed one into the other. When everything was stashed under cover elsewhere, we limped back to the barracks. Our feet hurt from slipping and sliding around in the mud; every muscle in our bodies screamed with fatigue. I made up my mind that if I were going to be a coolie, I'd have to brush up on the shifting-the-load technique.

The well was down on the other side of the camp, beyond the cement houses. We had to carry a bucket and rope for drawing the water, and by the time we had walked back to our shed in mud or dust, we needed another bath. The well house was much too small to accommodate all those who had to use it. It was a day of rejoicing when we saw workmen arriving to begin sinking wells in the areas between the barracks. At first the water was red from the clay, as the wells were not lined. But even this water was welcome for washing down the toilets. During the dry season, water became scarce and we were careful of every drop. During the rainy season, we could wash a considerable amount of mud and dust from our work clothes and legs by just standing in the rain. A bathhouse was constructed around the well site, and a cement guard wall was put in around the mouth of the well to keep the children from falling into it.

With soap—as with food, clothes, everything—sometimes there was just enough; other times the supply was far from adequate. If it were in short supply, we bathed with our clothes on to conserve the soap, for that precious item soaked through the dirty, sweat-filled garments and cleansed our skin as well.

We had two sets of clothes issued to us: shorts and sleeveless blouses for those of us under thirty; a dress could be requested by those over thirty. These garments were navy blue, but after the first wash they were less than navy blue, and after being in the sun for not too many days, they became gray. Because there were no lamps in the bathhouses, we bathed by braille, then put on our dry outfits for sleeping and were ready for work the following morning.

About two weeks after we had arrived from Malino, Mr. Yamaji was assigned to Kampili as permanent camp commander. Our isolation was so intense that very little of the outside world slipped across the moat or under the barbed-wire perimeter of our two-acre compound, but occasionally a rumor reached us. It was whispered that Mr. Yamaji had a maniacal temper and had just beaten to death an inmate of the men's camp at Pare Pare. We were dying for news of our own husbands, and those women who had sons also interned up the coast had double concern; but if the rumor were true, the new commander was certainly not the man to ask for that cherished information. We could only guess how our husbands must have felt—how helpless, how thwarted, how anguished—when they learned that this man was being transferred to Kampili as sole authority and arbiter over the lives of their wives and children.

One morning the camp was called to assemble before the Kampili headquarter office to meet Commander Yamaji. A short, fat man with bowlegs, he paced back and forth before those lines of women and children and a few remaining men. His face was round and moonlike, and his dark eyes watched from behind large, dark-rimmed glasses. His black hair was crew-cut, short, and spiky. When he issued commands, his parted lips revealed tobacco-stained teeth. He was dressed in a lightweight khaki uniform— short pants with knee socks and a singlet under his lightweight shirt. We were to learn that when his temper was aroused, he was like a man who had gone berserk; he could be deathly ruthless.

The second-in-command to Yamaji was an impudent young officer who bore the ill will of all. He strutted about and loved to remind us that we were his slaves; he often made an issue out of nothing. He was wholeheartedly abhorred.

Darlene and Russell in Brussels, Belgium holding church meetings during a weekend break in their language study in Holland.

Darlene as a young bride and a new missionary.

Darlene McIntosh and Russell Deibler on August 18, 1937, their wedding day, in Boone, Iowa.

The Deiblers' official work permit portrait, 1938.

Dr. Robert Alexander Jaffray arriving in Macassar on the day Darlene first met him in 1938.

A copy of the drawing of Russell's grave by Brother Geroldus given to Darlene by Ernie Presswood after the war. (This copy was drawn from the original by Bruce Rose, Darlene's son.)

Crossing one of the precarious
vine bridges constructed by
the Dyaks.

Darlene leaving the coast in a
canoe traveling toward
basecamp.

Russell with four young friends on the steps of the home he and the Dyaks built for Darlene.

The chieftain of the Zunggunaus from Zanepa bring their gift, a pig, to welcome the white women from the "Spirit World."

Meeting the first Kapaukus near Enarotali.

Darlene surrounded by the Kapauku women as they greet her, the first white woman they have ever seen.

One of the camping sites along the arduous trail from Oeta to Enarotali

Darlene in her first home with Russell. The walls were woven by the Kapaukus. Darlene carried the few decorations in her pack over the treacherous trail to Enarotali.

Darlene poses while holding fresh greens, a gift from Idantori and his youngest wife.

Kampili after the Allied
bombing on July 17,
1945. (All photos on this
page supplied by Elsie
David.)

The missionaries in Macassar, two months after their liberation. Despite the hearty post-war meals Lilian Marsh, Margaret Jaffray, Ernie Presswood, Ruth Presswood and Philoma Seeley remain severely under their pre-war weights.

Darlene Deibler, ten months after returning to the States. Her health remained precarious for years to come.

Beneath these two men, Mrs. Joustra continued to serve as the link in the chain of command between the women and children in the camp and their Japanese captors. She was thirty-eight. Large and angular, with skin that seemed to be permanently sunburned, she was gifted with unusual organizational abilities. She wore glasses with light plastic frames, and her hair was braided and wound around her head. Her work attire was a pair of shorts, pedal-pusher length, and a pair of heavy boots that amazingly lasted throughout the entire war years. She was a school teacher by profession, a fine woman who took her responsibilities seriously. As best she could, she juggled the insistent work quotas of the camp commander and the decreasing morale and physical condition of the women for whom she felt great responsibility.

Our days had been so busy since our arrival in Kampili that there had been little time to visit with Dr. Jaffray. I went over with Margaret every week to cut his hair and trim his beard and mustache, but I missed the visits and the Bible studies. One early evening in July, after I had finished my barbering, we had quite a lengthy chat. Dr. Jaffray loved people and had been relating anecdotes about those in his house on the other side of the camp. As Margaret and I rose from the bench to leave, he said, "Lassie, I love Mother and Muggie, but I'm so lonely for Russell." Two days later, the big trucks rolled into camp, and all the men (with the exception of a carpenter and Dr. Marseille) and the boys over sixteen were ordered to collect their baggage and board the trucks for deportation to Pare Pare. Dr. Jaffray was to get his wish; he would be with Russell.

From our barracks, Wang Mie Foo, Jan de Keizer, and Robbie Paul were to be taken. We hurried to say goodbye to Dr. Jaffray. He looked so regal with his white hair and close-cut beard and mustache. I felt like crying—must he also be taken away? That day we said farewell to a prophet, a great man of God, and a dear friend, and we all knew it. After they had loaded the men into the truck, Dr. Jaffray leaned toward me and said, "Lassie, whatever you do, be a good soldier for Jesus Christ." The echo of those words was to sustain me through the awful days ahead.

CHAPTER SIX

While darkness still shrouded the barracks, I listened to the stirrings of those about me. The grounds near the barracks had been completely denuded of trees, so there was never the cheerful twittering of birds to announce the new day and say, "Get up, you sleepyhead." I missed that.

The day before, Dr. Jaffray had been taken away, and my heart felt especially heavy: how grievous for Mrs. Jaffray and Margaret. "Lord, what will this day bring forth? Remember, You promised that 'as my day so shall my strength be.'"

I pulled my mosquito net out from around my mat and threw it up onto its ceiling, held taut by ties fastened around the bamboo poles at each corner of my rack. There was no need for dressing; I had slept in my clean work suit. Looking down the length of the barracks from my lofty perch, I could see that the "pot parade," as we called it, was under way. Never did we venture out to the latrines in the dark of the night without a stick and a prayer that no rabid dogs would be on the prowl.

Having refreshed myself at a tin of water set ready the night before, I picked up my tin plate and spoon to join those already in line to be served at the porridge drum. I ate hurriedly and washed my plate, setting it where the sun would later shine on and sterilize it. Then I ran with a few others from our barracks to the calisthenics class. We tried to have a fair representation of us younger ones, for we never knew when Commander Yamaji would appear in his singlet, shorts, and thongs to see if we were obeying his latest command. Calisthenics were mandatory, designed to make us fighting-fit to face the day—so he explained. They only made me fighting-mad to think of the needless expenditure of already dwindling strength. I was thankful when the commander finally wearied of emerging from his quarters to count us and do a few squats and bends before us, and calisthenics died a natural, long-overdue death.

Returning to the barracks, I was told that Mr. Yamaji wanted Rose David to report to the office immediately. "Good gracious, what does he want?" Rose exclaimed as she hurried toward the office.

I checked on the sick ones in our barracks and made work-detail substitutions. Coming back through the barracks, I realized that Rose had not yet returned. After the doctor made his rounds, Rose was *still* absent. I felt concerned. What now? Should I go to the office and check to see what had detained her? I prayed for wisdom and started across the camp toward the office. On my way I saw Mrs. Joustra's messenger and called to her: "Mrs. Voskuil, have you seen Rose David?"

Coming up to me, she casually put her arm through mine, turned me around, and started to walk me back to Barracks 8. "The commander is furious. The baggage of everyone who was taken to Pare Pare yesterday was examined. He found that Rose had sent a package to her husband with the boys from your barracks. He has imprisoned her in one of the small huts down by the edge of the camp past the piggery, saying she is to be hanged. The carpenter has been ordered to find rope and set up a scaffold. No one is to go near her. That's all I can tell you now."

I returned to the barracks feeling physically sick. I confided in Ruth, and we prayed for Rose's safety. It is true that they had said, "No letters." But nothing had been said about packages. Unfortunately, Rose had wrapped the socks she had knit for Carr and had written his name on the outside of the parcel. Even the sight of Rose's handwriting would have greatly cheered her husband, but that was not to be.

Elsie and the boys received the news quietly, and only by the clenching of their jaws and the pallor of their faces did they betray their agitation and concern. The afternoon was a long vigil; I tried to work on the road while keeping the office in sight. When the bell rang that we might return to our barracks, Rose had not been released, and Mr. Yamaji was nowhere to be seen. Supper was eaten quietly. Our emotions ran the gamut from anger at the Japanese, to anxiety for Rose, to frustration at our helplessness, to fear of further reprisals, to despair over the injustices we had already experienced, to a certain knowledge that our situation was not improving. We cried to God for help. Then, out of the experiences of a day that now seemed so remote, I heard Russell whisper, "Remember, dear, God said He would never, never leave us nor forsake us." Hope surfaced and I rose to go to Rose's boys, when someone cried out, "Rose is coming!" My impulse was to rush to her, but we all were schooling ourselves never to give the Japanese the satisfaction of knowing our fears and the distress they caused us.

At evening prayers our hearts beat as one as we voiced our thanks to God that Yamaji's anger had dissipated as the day wore on and Rose had been returned to us greatly shaken, weak from hunger, but unharmed.

The Japanese were fond of pork and the army of occupation needed a steady source of supply, so the piggery had top priority. A new structure was constructed of planks with cement floors. The pigs were given more consideration and had better accommodations than the women prisoners. Our dirt floors degenerated into mud floors during the rainy season and dust bowls in the dry time. However, for the sake of those who worked in the piggery, we were thankful. A well was sunk nearby, from which uncounted buckets of water were lugged daily for the scrubbing and rinsing of the cement floors to maintain a sanitary level of living for these animals. The camp commander shot dogs in the surrounding villages, which the women skinned, cut up, and cooked with the stems of the banana plant in preparing the hot meals for the pigs. Bamboo was gathered in a nearby grove for the cooking fires. It paid to have connections at the piggery—I could taste no difference in flavor between dog liver and calf liver. The commander got wind of the disposition of the dog livers and ordered it stopped— after all, the poor pigs needed the nourishment. The sows that were farrowing had to be kept under surveillance at night as well. Only the physically strong were able to cope with the long hours and heavy work required in the piggery, so it fell to the lot of the unmarried girls or young married women without children to be responsible for Kampili's prime industry.

After a stint at the piggery, I walked into our barracks kitchen to hear Commander Yamaji giving instructions for fireplaces on which we could set our open half-drums for boiling drinking water. I dared to suggest that if they would lay a whole drum on its side and cement it into the firebox, there would be a greater area of the drum exposed to the flame, thus requiring less bamboo to boil twice the amount of water. Further, the piece cut from the top of the exposed part of the drum could be flattened to serve as a lid to keep out the smoke. In my enthusiastic explanation of how it ought to be done, I hadn't realized how quiet he had become. I turned to get his reaction and saw his hand raised to strike. I caught the full force of the blow across my shoulder and the back of my neck. "You talk too much," he yelled and stomped off. It was a fight to keep the tears from spilling over—but the following day the workmen were cementing in the drums as I had suggested! The blow was a small price to pay for the added convenience and for drinking water free of the acrid bamboo smoke flavor we had always had in the open-topped half-drum.

Where you have pigs you have flies, and where you have flies to contaminate the food and water, dysentery can become pandemic seemingly overnight. It spread so rapidly that even the camp commander was quite shaken, because it was robbing him of his large work force. He decided that an isolation ward should be built for dysentery patients on the outer perimeter of the camp. Before the building, not unlike our barracks, was completed, he ordered the nurses to start shifting critically ill patients to the new structure. Elsie David, her friend Nita (a nurse's aide), and Sister Remolda (head of the hospital) had gone with Dr. Marseille to check the facilities. The bamboo beds were not in place and the toilets were filled with grass, bamboo curls, mud, and cement dust. It was far from ready for occupancy. Elsie and Nita started to clean the toilets while the doctor and Sister Remolda consulted with Commander Yamaji, who had suddenly appeared on the scene. He told Dr. Marseille to call the girls to start setting the bamboo beds where they belonged. Elsie heard Dr. Marseille call, but she whispered to Nita to pretend that they hadn't heard. They were anxious to finish cleaning the toilets before going inside. When the next summons was more insistent, they decided that they had better appear. Yamaji was in a terrible temper. "Did you hear us calling you?" he roared.

"Well, yes," Elsie answered, "but—" He cut her off, demanding to know why they hadn't come. "We were cleaning the toilets." His face was contorted with rage, and with his open hand he gave her a vicious slap across the side of her head. She was knocked to the ground, her head ringing. When she saw him grab his cane, she threw up her hands to protect her head and caught the full force of the blow across her wrists.

Sister Remolda tried to grab his arm, yelling, "Don't do that. She's only a girl." Infuriated, he started to use the cane on Sister Remolda.

Shocked that he dared treat a nun in that manner, Elsie tried to rise, yelling, "You can't hit her like that; she's a nun."

Further interference was more than Yamaji could endure. He started to kick Elsie with his heavy boot. She dropped back onto the ground, almost paralyzed from the pain near her kidneys. Yamaji must have thought that he had killed her, for he stopped dead-still, blinking his eyes, looking at her. Sister Remolda and Nita were on their knees beside her. A moan assured him that Elsie was still alive. He snapped, "Get this place cleaned up." With that he turned and left the building. It was well that Dr. Marseille had been shocked into immobility, for had he interfered, Yamaji would probably have killed him.

Without an X-ray the doctor could not ascertain whether any of Elsie's bones were broken, though he thought it possible that her eardrum had burst, for she was deaf in that ear. (Elsie's wrists were more seriously damaged than was thought at the time, for they remained swollen and healed deformed.) Her jaw had begun to swell and discolor. When Elsie could walk, Nita escorted her back to our barracks.

I was shocked when she stepped over the threshold. The left side of her face was swollen out of shape. "Elsie, you're hurt. What happened?"

Her grin was courageous and lopsided, but her eyes held none of her characteristic merry twinkle. I knew she was in pain. "Well, Mrs. Deibler, I just had a run-in with our good friend, Mr. Yamaji." Then she filled in the details, showing me her bruised body.

Rose had only to look at her daughter to know that something was terribly wrong. "Now, Mom, it's all right. Our good friend Yamaji just wanted to show us that he's the Big Boss." With considerable bravado, she told the story. To cloak her fear of what Yamaji might yet do to her daughter, Rose scolded her sharply—how could she be so stupid, knowing what kind of a man he was? Had she already forgotten what had happened to her mother?

Sensing how agitated her mother had become, Elsie tried to soothe her, admitting how stupid she had been and promising that in the future she wouldn't repeat the performance. Pulling the improvised curtains around their section, we left them to rest.

The bell signaled for us to stand up in the aisle at the foot of our beds and be counted. Having responded to our bows with a salute, Commander Yamaji, cane in hand, turned into our barracks and made his way to where the Davids stood before their beds. I followed him, as was customary when he made a bed-by-bed inspection, praying that the Lord would smite him with impotence to do further harm to this family. He stood and looked at Elsie, then at Rose. A hush pregnant with fear prevailed over the whole barracks. Finally, after what seemed to be an eternity, he shook his cane in Elsie's direction and said to Rose, "You have a very stupid daughter. I tell her to come and she doesn't obey. She's rude; she's stupid. Do you know that?"

In an attempt to pacify him, Rose immediately responded, "Oh, yes, sir, I already told her she was very stupid. I told her she should have obeyed you. Yes, sir, my daughter is very stupid."

Feeling somewhat vindicated, Yamaji turned to Elsie and, with a final shake of his cane in her face, said, "If I tell you to do something in the future, you obey. Do you hear?"

Her left eye now almost swollen shut, Elsie looked up at him very meekly, replying, "Yes, sir."

We made our way back to the front of the barracks. As he disappeared in the direction of the next four barracks, I released a sigh of thanksgiving that no more violence had been done.

In weariness of spirit and emotionally drained, I stretched out on my rack, reviewing what had happened, still seeing Elsie's battered body and bruised face. Phrases from the Gospel of Matthew were going through my mind: " 'Love your enemies.' 'Do good to those who despitefully use you.' 'Pray for your enemies.' All right, Lord, I'll pray for him. I sincerely don't want the man to be lost eternally—but I really would like it if You would curdle the food in his stomach tonight, and would You stretch him on the rack of his conscience—at least for awhile?" How very much easier it is to be philosophical about and forgive the wrongs done to oneself than to forgive the injustices done to the people we love.

With sufficient provocation, there is within each of us the potential to violence—but for the grace of Almighty God. "Without You, Lord, what might I have become? Forgive me, Lord." With a prayer for God to have mercy on the man, I drifted off to sleep.

The wave of dysentery had not abated, our isolation ward notwithstanding. It took more than a bit of soap and water to kill the dysentery germ. With little emetine for the treatment of the amoebic strain and sulfa unknown to us for the bacillary type, we had to trust that the patient was physically strong enough to be purged with salts repeatedly, then starved in an attempt to cleanse the system, stop the diarrhea, and allow the inflamed intestines to heal. When a measure of relief had been attained, rice water was given the patient, and later plain rice porridge. If the diarrhea remained under control, the patient graduated to plain rice in gradually increasing amounts. It was one thing to use such drastic treatment with adults who understand the why, but for the little ones, it was cruel. Sometimes it was almost not to be endured. O God, was it not awful—the pleading, hurt, haunted look in the eyes of the small ones, whimpering, not able to understand why they were being thus treated? The agony of the mothers, not being able to feed their sick babies, was almost unbearable. The eyes of the mothers took on the same haunted look as they clutched their little ones to their breast, rocking back and forth, crooning, and whispering words of love until sleep eased the stomach cramps and hunger pangs. I had to encourage the mothers to believe that their children never doubted that they were loved. When, like these little ones, I didn't understand God's dealing with me, I prayed, "Be patient, Father. Forgive my tears and whimperings. Just let me feel Your arms of love about me; then my heart shall understand. I shall stop my questioning and know that You mean it unto good."

The open drainage ditches carrying off the waste from the piggery provided ideal conditions in which the flies could breed and multiply. Even to the commander, the flies had become pestiferous, so something had to be done about it! There was no insecticide of any kind available. When he issued the order that every woman in the camp had to kill and personally deliver to his office 100 flies a day, we thought he had to be joking. He wasn't. He watched while we dumped our offerings into a tin set on the steps of his office. After some of the women got the cane treatment for falling short of their quota or for having cut the flies up to make it appear they had 100, we decided it wise to make every effort to collect the full amount. We never knew when we might be singled out to have our catch counted. The most flies were around the piggery, so whenever there was a free moment we raced off to the pig sties to make our kill. The sudden inundation of several hundred women, all intent on killing flies, must have been unpleasant for those working at the piggery, but we stopped only after an official edict announced that the practice *had* to stop, for it was disturbing the pigs.

We had to devise some other method, and someone remembered how attracted flies were to soapsuds, so before we started out to work in the morning, we whipped up a mound of soapsuds in a tin or on our plates and left them to lure the flies to their death. If by afternoon we were still short, we soaped our hands and went down by the latrines. Each slash of our open hand through a swarm netted us five or six. We didn't have time to let them die, so we squashed them between our fingers and, when our quota was filled, carried them to the office. We figured that approximately 60,000 dead flies were delivered daily to the commander, but there was no noticeable decrease in the fly population. It was a most irksome task. I suggested that the best solution to our fly problem was to consign Yamaji and his piggery to the dark side of the moon. Eventually Yamaji wearied of counting dead flies, and this, like the calisthenics class, died a long-awaited death.

One morning soon after the fly episode, I had several strips of gauze handed me with a note that these were to be made into face masks to be worn by the women who dished out our food. A larger piece of gauze to protect the food from the flies after it had been dished up in the central kitchen and while it was being carried in baskets to the barracks would have been more to my liking. But, fair enough, I thought; any precaution against spreading disease is acceptable. To say that the women breathing on our food were responsible for the increase in the number of sick people was, however, hardly a logical assumption.

We made and used the masks. The August heat was stifling and the masks were uncomfortable, so one morning one of the women slipped hers off and left it hanging on one ear. Of course, that had to be the day Yamaji sneaked in to make an inspection. He flew into a rage and confiscated all our food, although only a handful of people had been served. I had not yet joined the line of those waiting, so when they explained what had happened, I set my empty plate down and started for the office on a run to retrieve our dinner, lest in his anger he dump it. "He likes us to grovel, so Lord, help me to 'do a good grovel' for him." I arrived in time, for the baskets of food were sitting on his office floor. But when he suddenly appeared between me and the food, yelling, *Mau apa?* "Want what?" I almost forgot what I'd planned to say. I retreated to the next step down and looked up to explain how sorry the woman was she had taken her mask off. It was very wrong and it wouldn't happen again, but I had hungry children to be fed and hungry women who needed that food so that they could work hard that afternoon. They had done nothing wrong, so why should they be punished?

"*You* are the one who was wrong. You're head of the barracks; you're responsible to see that the women wear the masks."

"You're right, sir. I was the one who was wrong." I emphasized how *bodoh* I had been. Meaning "dumb" or "stupid," it was a word he loved to use when referring to us, so I pressed that point. That seemed to pacify him, and he motioned for me to take the food, then disappeared.

I motioned for some of the women who were watching to come help me carry the food back to the barracks, thinking as I left the office, "Lo, one woe is past and behold another cometh." The words were prophetic.

We had made lamps from small tins with a rag wick protruding through a floating lid, but the kerosene ration was so small that we disciplined ourselves in the use of them. However, after the grime of the day had been washed off, the announcements read, and devotions finished, most of us were ready to retire. Little visiting was done with friends from other barracks.

Since their move to Barracks 6 and after Dr. Jaffray's departure, we had not seen much of Mrs. Jaffray or Margaret. Occasionally Mrs. Jaffray helped prepare vegetables in the hospital kitchen, so Lilian was able to visit with her there. Margaret had joined the coolie gang unloading trucks—exhausting work—so we saw little of her. Margaret missed her father; the separation was very hard on her. This particular morning she had had an altercation with her mother, and, needing to give vent to her frustration and grief, she ran into the long grass behind the barracks and burst into tears. Suddenly she came face to face with Commander Yamaji, who demanded of her what she was doing there. She tried to explain, but he ordered her to get going and drove her before him to his headquarters, where he locked her in a room with a native man. She was accused of trying to make contact with this native man, whom Yamaji had apprehended earlier not far from where he had caught Margaret. Mr. Yamaji had been in hiding, waiting for the woman who was to rendezvous with the intruder. Fortunately, the man disclaimed any knowledge of Margaret; he had never seen her before and had not been trying to make contact with her. Realizing that the man was telling the truth and that Margaret had not gone into the long grass to meet him, Mr. Yamaji released her.

Margaret came to me as soon as he let her go and told me what had happened. "Thank God he believed you, Margaret. What about the man?" She didn't know, except that he had been tied up and left in the room. We talked until she stopped trembling and decided she had better go back to her work.

Suddenly the bell rang, and messengers were running to the various sections of the camp telling the prisoners to assemble immediately before the commander's office. Lilian Marsh came to stand next to me. When everyone had assembled, we were ordered to stand at attention. Then Yamaji appeared with the native man. He demanded of him the name of the woman, but the man only shook his head. Yamaji struck him, but he remained silent. The sun beat down upon us as Yamaji's eyes moved up and down the rows of women, looking for a face betraying guilt or fear. Our trial by ordeal had begun. We were ordered to form a line and, as we moved along, we were to stop before the native prisoner, turn and face him, then at a signal from Yamaji move off—all one thousand and more of us. The sun was at its zenith; the heat was oppressive. One row after another moved forward. I was convinced that the native man would never identify the person with whom he had been trying to make contact. Another row moved up. But would he falsely identify an innocent person? My skin felt clammy with fear.

Lilian was directly in front of me as our row started to move toward the prisoner. Her face was white as a sheet, and I kept worrying lest she faint from the heat and dreadful tension. One wrong move, a show of anxiety, an expression of fear, or even sympathy or exaggerated nonchalance—*anything* might trigger Yamaji into thinking you were the one. His eyes sought for betrayal in the face of the prisoner or the woman before him. If Lilian fainted . . . but she was a woman of great character and stamina. Ashen though she was, she stepped forward and faced the man and Yamaji without flinching, then moved off at the sign from Yamaji. My concern for her alleviated some of my own fear. I stepped into position, then was motioned away to follow Lilian back to our row.

The afternoon wore on until the last woman had returned to her place on the field. I realized that here was a person of noble character and unusual courage. The native man had refused knowledge of any of us.

Horrible as the ordeal had been thus far, it was nothing compared to what we were forced to witness as Yamaji unleashed his fury upon the man. The blows from Yamaji's cane were so vicious that bones must have been broken. The man's hands were tied behind him, so he could in no way protect himself. His teeth were clenched, but who could endure such punishment? At last an agonizing cry burst from his lips as he was pummeled to the ground. Throwing aside his cane, Yamaji began to kick the man with his heavy boots—the rib cage, the stomach, the head, no part of the body escaped. Dear God, will he never stop? The commander's insane.

Some closed their eyes, but it didn't close out the sound. I wanted to retch but was acutely conscious of the second-in-command, who stood to the side watching our reactions—when he could draw his eyes away from Yamaji and his victim. For the first time I understood "bloodlust"; it was reflected in a face alight with an unholy glee at seeing a fellow human being kicked and pummeled—perhaps to death.

Finally, exhausted, Yamaji stepped back. The man was still— too still. His clothing torn and soaked with blood, he lay inert and lifeless, no movement to betray whether he was dead or alive. He was dragged away; then we were dismissed. I felt emotionally ravaged and torn and prayed that I would never have to witness anything like that again. We never learned his fate or who he had attempted to contact—we never asked. But there were two things about the commander for which we came to have great respect—his cane and his boots.

To date, nothing had so completely demoralized us as the incident we had just witnessed, and we felt that measures must be taken to preclude its happening again. At a conclave of the barracks heads with Mrs. Joustra, we discussed the possibility of setting up our own guard system against these intruders—well-meaning though they might be. The women of the camp were in full agreement with our suggestion; but who would perform guard duty during the night? The barracks leaders volunteered a two-hour watch each, from seven at night until seven in the morning. This meant being on duty every other night. "And if we catch an intruder, what then?" It was agreed that we would report his presence to Mr. Yamaji. This scheme was devised as much for the protection of the natives as for ourselves. By the time we reached the office and returned to the fence with Yamaji, the native would be long gone.

My first night on duty was very dark. I stood still, allowing my eyes to grow accustomed to the faint outline of the barbed-wire fence silhouetted against the water of the moat. This was the start of my second turn around the perimeter of the camp. I clutched my stick and walked softly in bare feet, my ears straining to catch any movement in the grass that might mean a rabid dog. I was halfway to the gate on the east side of the camp when I froze. Something was moving the grass on the opposite side of the barbed wire. A dog? My hand tightened on the stick when a tall form rose up out of the grass. Every instinct to survive was sensitized by terror. I took a deep breath and let it out slowly, asking, "Who is it?" That's a futile question to ask any native, for the inevitable answer is, "It is I."

"Forget it, I don't really want to know your name, but I must report to the Japanese camp commander that I saw someone here. We're trying to protect ourselves and you. One of your people was badly beaten here last week. He may have died. Warn your people not to come here. Now go as quickly as you can and cover your tracks so you cannot be followed." He believed me, for he disappeared immediately into the surrounding darkness. I paused to gain composure, then walked slowly toward the headquarters building in the center of the camp.

It was around ten o'clock. Yamaji was not to be seen, but the odious second-in-command was there. I reported seeing someone outside the fence. As always his response was a very nasal, "Aw! Aw!" Maybe he understood Indonesian but didn't speak it.

Stepping out of the office to lead him to the place where the encounter had occurred, I realized that he hadn't followed me. Returning to the office, I found him buckling on his gun belt. He pulled out the pistol to make sure it was loaded, dropped it back into the holster, found a flashlight, checked it, then sat down to put on his heavy boots. Motioning that I should lead out, he now followed down the steps. All the while flashing his powerful beam of light in every direction, he remained at my heels until we reached the barbed wire. I had known he was a poor specimen but would never have credited him with such cowardice. He kept me in front of him all the way to the fence—just as I used to herd my younger brother, Ray, ahead of me down the dark cellar steps. Did he think he was in danger from the native, or was he afraid that I might jump him from behind? The daytime intimidator of unprotected women was a nighttime coward—loaded gun, flashlight, and heavy boots notwithstanding. We women walked around barefoot and alone, even during the darkest hours of the night, around the outer areas of the camp, armed only with a stick as a defense against mad dogs—and the contrast angered me. To be afraid is nothing to be ashamed of, but cowardice, that's something else! I hated going down beyond the piggery, feeling along the fence in stygian darkness. Invariably the oppressive stillness would explode into a cacophony of shrieks and pig squeals, and you never knew what or who startled the beasts. I disliked it, for I'm not a brave person; but I never skipped the area or ran.

I indicated the place where the person had been. Except where the grass was still depressed, there was no evidence of anyone having been in the area. I could scarcely guess which of us was the more relieved. The flashlight beam probed the darkness around the area. The investigation was over and he was starting back toward the office; now I was in the position of rearguard covering his retreat. I thought it would be a pity if he had not one memento of his brave search for the intruder, so I stepped out to the side of him, saying, "This way, sir," and led him through a beautiful field of sticktights. By the time we reached the area of light near the office steps, his immaculate, three-quarter white socks looked like furry gray puttees. He looked down at them in dismay. When I saw his expression changing to anger, I bowed hastily and said, "Good-night, sir," then beat a retreat, remembering the quip that cowards live to fight another day. All the way back to the barracks, I thought about the fun he would have removing those sticktights. From the malevolent looks he cast in my direction, I had a feeling he didn't care much for me after that.

Reviewing the whole incident later, I wondered as to the where-abouts of the commander that night. Had he been stalking me? Could it be that he had witnessed the bizarre display of cowardice perpetrated by his underling? Or was it just that Yamaji found a daily diet of "Smarty's" buffoonery more than he could stomach? Whatever the reason—the second-in-command was mustered out of camp a few days later, not to return.

In a day or two, Smarty's replacement arrived. I don't know what we had expected, but whatever it was, this wasn't it. The car door opened and out stepped a Japanese lad who couldn't have reached his twentieth year. He was as skinny as any of us POWs. His neck was long and scrawny, and his head bullet-shaped. He was wearing a regulation-type Japanese army uniform. When Com-mander Yamaji appeared, the new man doffed his cap, bowed, then stepped forward with a smile and a greeting to shake hands. We could see that his hair was clipped in a crew-cut not unlike Mr. Yamaji's. Then, in acknowledgment of the women witnessing his arrival, he turned a pleasant, boyish face toward us and with a bright smile bowed several times in our direction. We responded with a feeling of warmth and relief. He was promptly dubbed "Sweet Seventeen." Not until he appeared the following day in the casual attire of singlet and knee-length shorts did we realize what big feet he had. Their size was accentuated by his ankle-high boots and the khaki socks that drooped down over the tops of the boots. When he walked, he leaned forward like a person on the edge of a swimming pool ready to take a plunge. That Commander Yamaji was happy with him as his assistant was evident.

At roll call that evening, Sweet Seventeen accompanied the commander and Mrs. Joustra. After we had bowed and greeted them, Mr. Yamaji said something to him in Japanese. I caught two words of the conversation—"America" and "Deibler." Off came

Sweet Seventeen's cap, and he bowed to me the second time; then they continued on to the remaining barracks.

Before long I realized that I had been singled out: no matter where my work took me in the camp, Sweet Seventeen would turn up, smile, doff his cap, and greet me, bowing very low. He never said anything other than a greeting, but I became embarrassed when other women noticed and remarked about it. I solicited their help to warn me of his presence in the vicinity; then I would disappear. However, he still seemed to find me. There was nothing in his conduct toward anyone in the camp that was not exemplary and kind. Nevertheless, I felt an increasing uneasiness.

To counterbalance the tragic, we were always ready to see the ridiculous or funny side of any happening in the camp and to respond with a good laugh. There were many women whom we never learned to know except by sight. During one particular period of a week or ten days, very casual acquaintances from other barracks sauntered by to make small talk with me. They often threw in a personal question or two of such a ridiculous nature that I responded with a vague or flippant answer. Thinking about the strangeness of these questions, I began to see three categories emerge—my father's money, films, and flying.

One morning I was visited by a woman of the former upper echelon of colonial society, whom the hardships of imprisonments had unmasked to reveal her for what she really was—a greedy, conniving woman. Evidently she felt it her prerogative to find out the truth about me. She faced me that morning, arms akimbo like a "fish wife," as the Dutch would say, and in a very insolent tone inquired, "Is it true what the boys are saying about you?"

"What boys? And what are they saying about me?"

"You know, the boys from your barracks. They say you're from a very wealthy Texas family, you're a movie actress, and you fly your own plane."

Her denigrating tone when mentioning the boys from my barracks raised my hackles. "Really, Mrs. Belksma, you know how boys talk!" I responded, turned on my heel, and walked away. Two things were clear: I wasn't betraying the boys to a woman of her ilk; second, I needed an early conference with the boys.

That afternoon when the teenagers returned from work, they were chattering and laughing. Seeing me, they called, "Good afternoon, *Mevrouw* Deibler."

"Good afternoon. Oh, by the way, I have a problem. As soon as you boys are bathed, would you meet me on the west side of the barracks? I'd like to talk to you."

Shocked silence followed. Someone said, "Yes, *Mevrouw* Deibler." They formed into a tight huddle as they moved off down the center of the barracks, whispering to one another.

They were squatting in a circle arguing when I joined them. They started to rise, but I motioned for them to remain seated and found a place for myself. It was difficult to maintain a sober countenance; they looked so ill at ease. I gave them a quick rundown of questions various women had asked, then of the visitor I had had that morning. "So, boys, would you like to tell me what this is all about? If I'm supposed to be a millionairess, an actress, and an aviatrix, maybe it would be well for me to know how I became all these things."

They looked at one another for a second; then Normie, who evidently had been delegated spokesman, explained how I came to find myself suspected of being a member of the moneyed class. I visualized the whole drama as Normie told it (with verbal jabs from one or the other of them). It went something like this: The young people had been assigned to clean the area on the east side of the camp near the fence. When they all stopped for a breather, in lieu of "my father can beat up your father," they substituted barracks leaders! One of the lads from another barracks suggested that the rest of the leaders were a bunch of commoners. "Yeah," added one of his friends. "Our barracks leader is the wife of the *burgermeister* [mayor] of Macassar."

"Burgermeister? Ha! Our leader's husband was patrol officer over a large area in the Lesser Sundas."

"Burgermeister? Patrol officer? That's nothing. Our leader is the wife of a governor."

I could imagine the wheels turning at top speed in the minds of the Barracks 8 boys. They, the foreigners, really had to come up with a story that was the granddaddy of them all. They put their heads together in a whispered conference. Who suggested what, no one could recall, but when the others had exhausted their résumé of the importance of their leaders, my kids were ready to leave them without a feather to fly with. It was impossible to keep the mirth out of my eyes or keep the corners of my mouth from twitching. Realizing that I wasn't angry, they all chipped in, lest any detail be omitted.

"Well, what's so great about all that?" one Barracks 8 boy had asked. "Are *they* the burgermeisters or patrol officers or governors? They're just the wives. What have they done? Sure, your teacher has had a lot of education and she has brains, but the camp's full of teachers. We don't know if we should tell this, because we don't think she really wants people to know—you know how it is with movie stars!"

"A movie star? You mean Mrs. Deibler is a movie star?"

"Ask her—no, you'd better not or you'll get us into trouble. You see, she comes from a very wealthy Texas family. Oil, we think, but whatever—they have scads of money. She has her own plane in America. Her dad gave it to her and she does her own flying, learned when she was very young—probably fifteen or sixteen, because she's only in her twenties now."

So that's why I had been getting the once-over from the teenagers and why the women had been asking questions. A good thing that vitamin deficiency hadn't spoiled my clear complexion yet! Recalling some of my innocent, flippant retorts, I knew that they could have been misconstrued as evasive answers to hide the fact that I really was a wealthy actress trying to remain incognito. And to think I didn't even have a pair of dark glasses! I had to tell the boys about the woman who asked in an offhanded way if my father was in oil. "Oh, no, my father's in lumber," I said. (He worked at the saw mill in the railroad roundhouse.) Had I done any flying? another woman asked. "Only a little—nothing bigger than a twin-engine plane here in Macassar." (How was I to know that I was supposed to be a pilot?) Was I expecting to do a film on my war experiences when I got back to the States? I wondered if they thought all Americans were rich and lived in Hollywood, but I answered, "Probably. I'll write the scenario myself and title it *Riches to Rags in Kampili.*" It was so ludicrous that we laughed until we were weak.

"*Mevrouw* Deibler, those were *prima* answers," Bertie said. "And you're not angry? You won't give us away?"

"No, I'm not angry and I'll not give you away, but I'll not lie to anyone. However, this you must promise me: you'll not embellish the story any more. Promise?"

"Promise!" I stuck out my right hand and nine right hands closed over mine. We had a pact. I stood to break up the confab when one of them quietly assured me, "*Mevrouw* Deibler, we really do think you look like a film star."

Tears welled up, "Thanks, boys. I know what you're trying to say, and I think you're the greatest!" They made my day, for I felt it was the beauty of my Lord they had seen. I knew I had established a valuable rapport with the greatest group of teenagers in the camp.

With the coming of women and children from Bali, Lombok, Sumbawa, and adjacent smaller islands, empty beds were sought. Sparsely filled Barracks 7 received the bulk of the newcomers. Then I was told to make a bed available for a woman who was to arrive in a day or two. No further information was forthcoming as to who she was and why she was to be put in Barracks 8 when there were beds free elsewhere.

I protested that we had no empty beds, but the camp commander pointed out that an upper rack could be made available midway in the barracks by placing two smaller children in one bed. Apart from isolating the foreigners to Barracks 8, the Japanese had left the housing of the prisoners to the barracks leaders and Mrs. Joustra. Why was the placement of this woman in our barracks of special interest to the Japanese? The more we speculated on the whys and wherefores, the more apprehensive we became. Then she arrived—a veritable Amazon, almost six feet tall, weighing well over two hundred pounds, and with the abounding energy and good health to which we had become strangers. It was difficult not to be suspicious of those who were *gemuk basah* ("wet fat"), as the Indonesians say, in contrast to the rest of us, who had become *kurus kering* ("dry skinny"). A prisoner from the Lesser Sundas warned us to be extremely careful, as this woman was suspected of being a spy. Committing my people to God's protective care, I prayed that He would completely frustrate whatever purpose the enemy had in placing this woman among us.

The heat of the afternoon was oppressive. Tiele Noll and I had just pumped the vats for the camp kitchen full when the jangle of the bell signaled that we should assemble before the commander's office. The women from the piggery, garden, coolie, and kitchen crews were told to return to their work while the rest of us lined up by barracks. Mr. Yamaji then walked up and down between the rows, looking us over from top to toe like a stockman selecting prize cattle. Some of the women were told to move to the fore. Ruth Presswood was one of those chosen. He then explained that since dysentery had taken its toll of workers at the piggery, replacements were needed. The newly chosen recruits were to report to work the following day.

Returning to the barracks, I found Ruth in tears. "Darlene, if I have to go down there to work, I know it will be the death of me." I prayed silently. "O God, another crisis—what are we going to do?" Her cough had become persistent and of late she had been bringing up blood, so I knew her statement was no morbid, idle pronouncement. "All right, Ruth, you pray, and I'll go talk to Yamaji." I found him in his office and explained Ruth's physical condition to him.

"I understand, but you must find the substitute for her from your barracks."

A substitute? The only two young women who might be strong enough to do the work had small babies. It would be unconscionable to expose those little ones to the swarms of flies and the heat of the pig sties. Then I had a sudden inspiration—the Amazon! "Lord, I don't even know where she's been the last few days, but she would be the perfect substitute. Please make her amenable to taking Ruth's place." Before I could ask her, the Amazon came to me after evening roll call to say she had heard I was looking for someone to take Mrs. Presswood's place at the piggery and she would be happy to do so. I was very grateful that she had offered and expressed my gratitude. Who had told her? Yamaji? One thing was certain: God had answered prayer. The verse came to me, "They meant it unto evil—putting her in our barracks—but God meant it unto good."

Ruth was finding the duties of barracks nurse very difficult. She was an RN accustomed to working under sterile conditions. It drove her up the wall to see a hundred people or more being given injections with the same needle, without it ever being sterilized between shots. "There are diseases that can be transmitted from one person to another by such a careless practice!" she objected. I asked Dr. Marseille if he thought I could function as barracks nurse. He assured me that he could teach me a thing or two—thus I became barracks nurse; and even though Ruth went to work in the sewing room, she was always available to answer questions and make suggestions. At the war's end, a physical examination revealed scar tissue on her lungs, so her assumption as to the seriousness of her condition had been correct.

September brought a bright new personality to Kampili— Father Bell, an American Catholic priest who was by avocation a butcher. His skill was needed in slaughtering the pigs for the Japanese army in Macassar. The pig's blood was caught and made into blood pudding to be given to those in our camp suffering from severe anemia. The tripe was the best thing that had happened to our diet in a long time!

The truck that brought Father Bell from Pare Pare also delivered several large gunny sacks of wooden clogs. These had been made by the men of that camp for their wives and daughters in Kampili. Ruth and I each received a pair. They were beautifully shaped and polished. I recognized the material from Russell's shirt, which had been used to cover the straps cut from an inner-tube. It was very exciting and comforting to receive these gifts of love from our husbands, for it was now more than a year since we had had word from them.

Whenever Father Bell was free, he was besieged by women asking about their husbands, sons, and fathers. Many times, seeing Father Bell surrounded by a group of women, I moved to join them, but the Lord always checked me, saying, "Not now, you can trust Me. They need reassurance." Peace and quietness filled my heart as I obeyed the Lord and returned to my barracks.

Kampili lay between two major airfields, now under Japanese control. Zeros came and went from raids. We never knew their objectives, but one noon a Zero returned with an Allied fighter plane hot on his tail. A dogfight took place over Mandai, the airfield to the south and east of camp. The Zero dived, rolled, and tried to shake the Allied fighter, but without success. Time and place ceased to be. We had escaped into a world outside our prison. We were up there with the Allied pilot, praying for him, intrigued by his every strategy to bring down an enemy instrument of war.

Suddenly, after a burst of machine-gun fire, we saw the Zero explode into flames. The pilot bailed out before the Japanese plane crashed into the ground. "Hoorah!" erupted from a thousand throats. In our total absorption, we had failed to notice that Mr. Yamaji had moved in among us. As the Allied fighter disappeared in the distance, an enthusiastic woman continued to yell, "Hoorah! Hoorah! He shot him!"

Like a man possessed, Yamaji brought her cheer to a painful end. He flailed her with his cane until her screams of pain were unbearable. He raged that that was one of his people—he had feelings. At last, exhausted, he ordered us all back to work. As unobtrusively as possible, we put distance between us and him. We recognized our need to be more cautious in the future. We further knew that his recent order to dig slit-trenches was a wise measure. We needed someplace to shelter when the conflict moved into our area, involving these airfields. We would have liked bomb shelters, but only one was constructed, and that was exclusively for the Japanese officers, so we deepened our trenches for greater protection.

We were also instructed in air-raid procedure. As soon as planes were heard or seen, the bell should be rung. All lights must be doused immediately and every barracks and house evacuated. The first night the air-raid alarm sounded, we ran to our assigned trenches, helped the elderly, and accepted the children who were handed down to us. Lying on our backs, we searched the sky for planes. Only when they were caught in the searchlight beams could we locate them—bombers moving steadily toward the airfield nearest us. Our hearts were filled with terror, lest some of the planes drop their deadly shrapnel bombs before they had cleared the camp area. The ground shook; the stillness of the night was shattered by the reverberations from the exploding bombs. In the midst of the confusion, I could hear the Lord saying over and over to me, "It shall not come nigh thee. It shall not come nigh thee. It shall not come nigh thee." I knew that this was a portion of the ninety-first Psalm, but search my mind as I might, I couldn't recall the rest of the verse.

The planes, having released their bombs, circled and winged away into the east. All-clear was sounded, and we returned unharmed to our barracks. "It shall not come nigh thee." I could hardly wait for morning to read the rest of the verse. With first light I grabbed my Bible and read, "A thousand shall fall at thy side, and ten thousand at thy right hand; but it shall not come nigh thee" (v. 7). What a precious promise to carry in my heart during the many long, terrifying nights to be spent in the open slit-trench. There were many occasions when I might have questioned whether those words really meant what they said, but I steadfastly reminded myself that this was God's promise to me; I would believe Him. "It shall not come nigh me."

When the southwest monsoon began to blow across the island, bringing with it torrential rains, the ditches filled with water. The ground drank thirstily until it was sated; puddles and mud remained. Nevertheless Yamaji was adamant: no one might remain in the barracks during air raids. We were instructed to ensure that all our people headed for the ditches as soon as the alarm was sounded, by day or night, in rain or dry.

One night, in the middle of an alert, the commander checked several of the barracks and discovered that a number of mothers had remained inside with their children. It mattered not to him that it was drizzling rain and some of the children were ill. His orders had not been obeyed—that he could not tolerate.

As soon as the all-clear was sounded, the barracks leaders were summoned to meet Mr. Yamaji in front of his office. The rain had stopped and the sky had cleared. The starlight revealed a very agitated man. He walked restlessly back and forth, slapping one boot with his cane, waiting for all of us to gather. We were told to form a semicircle around him; then he began questioning the leader of Barracks 7. "Did all your women go outside?" he demanded.

"Well, no, sir, but . . . "; then she began to explain why the women had stayed behind, excusing and justifying their behavior because of the sick children and the rain. That this woman dared to defend those who had defied Yamaji's orders was more than he could countenance. In two angry strides, he was within reach of her. He struck her with his open hand on the left side of her face with such brute force that she was propelled through the air to land about ten feet from where she had been standing. The braid of her long gray hair, which she wore wrapped around her head and secured with long hairpins, was lying loose in the dirt beside her. My first thought: a blow violent enough to throw her that distance and loosen her hairpins has certainly broken her neck. I felt physically ill. But unaided she finally stood to her feet and returned to her place in line, her face badly bruised and swelling.

He approached me next. "Did all your women go outside?" he roared. As quietly as possible, I replied that everyone had been told to go outside, that the bell was ringing. Fully expecting a like blow, I braced myself, but he went on down the line questioning the others. We were finally dismissed with a reminder of what would happen the next time there were further infractions of the rules.

A look at the Barracks 7 leader's face was sufficient warning for most. But there are always those who think, "It can't happen to me. I'll never get caught!" Not too many days later we had an air raid at noon. Most of us had finished eating and were getting ready to go back to work. We dropped everything and ran for the slit-trenches. We were concentrating on the squadron of planes in the distance, wondering what was being bombed, when we heard someone roaring in pain like a wounded bull. I stood up to locate the sound and saw the commander driving Lena Paul before him. With every step he dealt her a vicious blow with his cane. She was so badly bruised that to sit was excruciatingly painful, and she had to sleep on her stomach for weeks. I never again had to check to see if we were all in the ditch.

CHAPTER SEVEN

One Sunday evening near the end of November, Father Bell was holding mass for those of the Catholic faith. At the close of the service, he said to those who had gathered, "There is someone in this camp who needs prayer. I can't tell you what it is, but if you know how to pray, pray for her." There was much speculation that night as to the identity of the woman to whom he referred.

The following morning, Mrs. Joustra came to our barracks and asked if she could talk to me for a few minutes. We walked out to the grassy plot beyond the dining area. As we paced back and forth, we discussed the work and how those of us who were younger could substitute for the many who were ill, sometimes doing two or three extra jobs a day in order to fill the work quota. Suddenly she stopped and said, "But I didn't really come to talk to you about the work. I came to tell you that your husband in Pare Pare has been very ill."

When I saw her eyes fill with tears, I grabbed her shoulders and cried, "Oh, Mrs. Joustra! You don't mean he's gone!"

"Yes," she said. "Some three months ago he died in the camp in Pare Pare."

I was stunned—Russell is dead. He'd been dead three months already! It was one of those moments when I felt that the Lord had left me; He had forsaken me. My whole world fell apart. I walked away from Mrs. Joustra. In my anguish of soul, I looked up. My Lord was there, and I cried out, "But God . . . !"

Immediately He answered, "My child, did I not say that when thou passest through the waters I would be with thee, and through the floods, they would not overflow thee?"

Turning back to Mrs. Joustra, I nodded mutely and walked back to the barracks. I grasped my ladder and rested my forehead against one of the rungs. I closed my eyes, trying to assimilate the awful, cold hopelessness within me—three months, Russell had already been dead three months; he had already been buried! The

terrible finality of it overwhelmed me. I was too new to such grief
to know how to handle it.

I lifted my head and saw the living who needed me, who were
waiting for my ministrations as barracks nurse. Beyond feeling, I
moved through the barracks. There was also a water vat to be
filled for the central kitchen. Starting toward the pump, I looked
up and saw Sweet Seventeen headed in my direction. I stopped
and waited for him to come up to me. As always, he doffed his
cap, bowed very low, then straightened up and awkwardly shook
my hand. I realized suddenly that he had known of Russell's death,
that all the greetings and bowings were a seventeen-year-old Japanese boy's way of expressing sympathy—a gesture of compassion
from a young lad who probably did not want war any more than
we did. As on another day—that Friday, the 13th of March, when
Russell was taken away—profound quietness began to seep into
the crevices of my broken heart.

By noon the news of Russell's death had reached into every
corner of the camp. All the afternoon and throughout the evening,
friends stopped by to express sympathy by a pressure of the hand,
a look, tears, or a few well-spoken words. Father Bell came to
explain why he had never told me of Russell's death. The Japanese
had bound him to secrecy by a threat of death. Surreptitiously he
had learned my identity from the Mother Superior and tried to
avoid me. "I decided that if you should come to inquire about your
husband's well-being, I would tell you that Russell was very happy,
for I knew he was."

Then I knew why the Lord had checked me each time I had
planned to talk to him. I would have made it very difficult for the
priest; possibly I would even have endangered his life.

The previous Sunday evening following mass, the priest had
been informed by Mr. Yamaji that the authorities in Macassar had
finally granted permission for me to be notified—was that not an
almost immediate answer to prayer? He sought Mrs. Joustra's help
in breaking the sad news to me.

With further comments on the friendship and mutual respect
that had existed between Russell and him, "though we were of
different faiths," Father Bell took his leave. I was left alone again
to grieve. An elderly man who had no known relatives in the
Indies had died previously, and another man had been beaten to

death by Mr. Yamaji, but Russell was the first of those who had a family in Kampili to die in Pare Pare. Back in the barracks, I gently handled the wooden clogs. Had Russell made them, or someone else—Ernie Presswood?

Late that afternoon, Mr. Yamaji called me to his office. He was standing behind his desk. "*Njonja* Deibler, I want to talk with you," he began. "This is war."

"Yes, Mr. Yamaji, I understand that."

"What you heard today, women in Japan have heard."

"Yes, sir, I understand that, too."

"You are very young. Someday the war will be over and you can go back to America. You can go dancing, go to the theater, marry again, and forget these awful days. You have been a great help to the other women in the camp. I ask of you, don't lose your smile."

"Mr. Yamaji, may I have permission to talk to you?" He nodded, sat down, then motioned for me to take the other chair.

"Mr. Yamaji, I don't sorrow like people who have no hope. I want to tell you about Someone of Whom you may never have heard. I learned about Him when I was a little girl in Sunday School back in Boone, Iowa, in America. His name is Jesus. He's the Son of Almighty God, the Creator of heaven and earth." God opened the most wonderful opportunity to lay the plan of salvation before the Japanese camp commander. Tears started to course down his cheeks. "He died for you, Mr. Yamaji, and He puts love in our hearts—even for those who are our enemies. That's why I don't hate you, Mr. Yamaji. Maybe God brought me to this place and this time to tell you He loves you."

With tears running down his cheeks, he rose hastily and went into his bedroom, closing the door. I could hear him blowing his nose and knew he was still crying. We weren't supposed to leave the presence of a Japanese officer without permission; however, since he didn't return to dismiss me, I sat quietly praying for his salvation, that he might understand new life in Christ Jesus and someday go home to share God's love with his wife and family—to be a light in some dark, possibly even remote, area of Japan. Realizing finally that he was not coming out of his room, I left, knowing from that moment on that Mr. Yamaji trusted me and understood why I was in the Netherlands East Indies. How adequate his Indonesian was to fully understand what I shared with him, I didn't know, but there had definitely been a response.

Roll call and devotions over, the women moved back to their beds. They whispered or spoke in muted voices; even passers-by talked softly. I said goodnight and climbed up to my rack. When I stretched out face down on my mat, I wanted nothing so much as a shoulder on which to put my aching head, and to sob until the fountain of my tears ran dry. I felt vulnerable and young, desperately needing the strong, comforting arm of the Shepherd. Who can bruise and make whole again? Who can break, then restore that which is shattered to a thing of beauty?

Suddenly my Lord was there, standing in the cathedral of my heart, and from His Word written upon the scroll of my memory, He began to read, "He hath sent me to bind up the brokenhearted . . . to comfort all that mourn . . . to give unto them beauty for ashes, the oil of joy for mourning, the garment of praise for the spirit of heaviness" (Isaiah 61:1–3).

From force of habit, I "spread it out before the Lord." Never once did He interrupt while I told Him everything about the past, the present, the timing, what I was feeling—or not feeling—how the future looked, and the oppressive feeling of aloneness. I waited to hear what my Lord would say, and silence answered me. How unusually quiet it was in the barracks that night. Then, like seeing clearly for the first time what the Artist had in mind, I understood—these people shared my grief. By their very quietness they were saying, "We're thinking of you and praying for you in your time of sorrow." Their words of sympathy, their tears, the warm pressure of their handclasps were the delicate shadings of empathy mingled with the warm hues of love to create upon the canvas of my heart a beautiful picture of friendship. "Thank you, Father, for these my friends. Thank you so much for these beautiful people, who are so very dear to me. Let my grief be mine alone. Anoint my countenance with the oil of joy, that none may ever feel embarrassed to laugh in my presence. May no joke or sharing of the ridiculous be stifled because I am there. Wrap me in the garment of praise, that I may not burden others with the heaviness of my grief."

Experientially, I was learning to understand the comfort of the Holy Spirit. Sometime during the dark hours I slept. The sword of sorrow had pierced deep within me, but He had bathed the sword in oil.

Before the year's end, a third of the camp was ill with dysentery. The isolation barracks and hospital were filled to capacity. Only the most severe cases were hospitalized. Those in the last stages of the disease were carefully guarded during the night hours to keep rats off their beds. As their very life flowed from them in a diarrhetic discharge of blood and mucous, the rats—vile, despicable creatures—came smelling death. I had not functioned long as barracks nurse before I could recognize bacillary dysentery by its smell. The proper drugs for treating the disease were nonexistent. Lacking disinfectants, we had to rely on the sun, when it shone, to sterilize the things used by patients.

Worms, ugh! Roundworms, hookworms, pinworms—we had infestations of all of them. Any treatment for them took the form of folk medicine. If we had had enough garlic, we might have used it for pinworms—but we didn't. There was nothing for hookworms. The roundworms—dreadful! We experienced them crawling out of the noses or mouths of people while they slept. I remember one missionary who nearly choked when one got up into her throat. Fortunately she was able to get hold of it and pull it out. Children with bad infestations had bloated stomachs. Often the toddlers would waddle around the barracks area or outside and, before they could make it to a container or toilet, would have bowel movements filled with writhing, long, white worms. If the mother of the little culprit were to be found, she cleaned it up; if not, the barracks nurse—you know who—did it!

Mrs. van der Kley became quite ill, and her symptoms puzzled the doctor. One morning she was unable to report for work. As I was making my rounds, checking on those who remained behind, she suddenly yelled, "*Mevrouw* Deibler, come quickly! I'm sick to my stomach!" I grabbed a basin and dashed over to her. She was violently ill and began to vomit. I felt quite nauseated myself when I heard things "plinking" into the basin and looked down to see about a dozen nickel-sized worms floating around. They were curled up like a large lima bean that was germinating. Her problem was quite evident, so I gave her the "red-hot-onion cure." I collected six or eight small red onions at the central kitchen, peeled them, and watched while my patient ate them. They were hot enough to burn up most anything they came in contact with. Then she "enjoyed" a massive dose of Epsom salts.

We begrudged the use of kerosene for anything but our tiny lamps. Sometimes, however, it was necessary to prescribe a "kerosene cap." Esther Meshmoor came to me half-crazy from head lice. "*Mevrouw* Deibler, could I please get some kerosene?" Her hair was covered with nits, so I cut it all off and burned it. Her mother, Rachel, killed the lice that tried to escape; then I applied kerosene lightly to the scalp. Esther slept with a rag tied around her head. The next morning we washed her head. Careful though we had been, she had a blister on the top of each ear from the kerosene that had seeped through the rag. Her gratitude to be free of the crawling lice was ample reward.

Kampili was visited by a team of Japanese doctors doing research on the effects of vitamin deficiency. Everyone in camp had to submit to a physical examination. We proved to be a most interesting assortment of guinea pigs. We benefited in no way by their findings, however. No vitamins were ever supplied to augment our diet, so we continued to resort to whatever medicinal herbs and roots were cultivated by those who understood the use of "green medicine."

A drink made from cooking the new leaves at the very top of the papaya tree was given those with malaria. We pounded rice husks in our large wooden mortars to coarse flour for making bread to be given to those who suffered vitamin B-deficiency-related diseases, such as the beriberi from which I was suffering. This type of roughage was a good source of vitamin B, but extremely hard to digest.

We made rice powder to dry the skin eruptions, rashes, and eczema. We seemed to find no source of vitamin A for those who suffered from night blindness, nor was there help for the mentally afflicted.

One of the Armenian boys, Dougie Marcar, had a severe case of tonsillitis. Being barracks nurse, I took him to the clinic. The Dutch woman doctor decided that it was imperative his tonsils be removed. No sooner decided than he was seated in a chair with me holding his head. There were no clamps or devices for sucking up the flowing blood, so I sopped up the blood in his throat with swabs after Dr. Goedbloed reached in with a wire instrument and removed the diseased tissue. He was a very brave young man. I walked him back to the barracks. After a few days in bed, he made a full recovery.

With only local anesthesia, an emergency abdominal surgery was performed on one of the nuns with complete success.

Rabid dogs became an enemy more insidious than the Japanese. We took to the ditches armed with homemade spears and knives for protection. By day or by night, they intruded into the camp. Just a nip on the elbow from a mad dog, and a little boy from Barracks 7 died a violent, painful death. I learned that when the stars were our only source of light, I could distinguish anything moving over the ground by standing in the trench with my eyes at ground level. My spear had a knife lashed to the end of it. Some of the boys from our barracks had fashioned it for me. Only God knows how many nights I watched and reminded Him of the Psalm that says, "Thou shalt not be afraid of the terror by night."

True character surfaces when people are faced with danger and deprivation. We had base characters among us—some who surprised us by their selfishness and greed—but they were in the minority. They were far outnumbered by those of great moral fiber, noble characters whose courage, fortitude, and valor were an inspiration to all.

Such a one was Toos Hoogeveen, a beautiful young woman, the married daughter of the former governor of Bali. Mrs. Hoogeveen worked at the piggery. One night when the air-raid alert sounded, she and some of the other workers took shelter in a nearby trench. They became aware of the approach of a dog and tried to ward it off with sticks, but it refused to go away. Realizing that the dog was rabid and many would die horrible deaths if it got into the trench, Mrs. Hoogeveen wrestled the struggling animal to the ground and held it until another girl killed it with a knife. In attacking the animal with her bare hands, she was bitten and scratched repeatedly. The Japanese responded to the urgent plea for help by sending some antirabies serum, but for some reason it was not effective. The hopeless desperation of the doctors was like a disease that infected the rest of the camp. What was to be done for her? Understanding that the serum had not effected a cure, she asked and needed to be isolated. The door of the hut was closed; the key was turned in the lock—and there she awaited death.

When it came, each of us died a little. Here was one who wore courage like a bright, shining badge. Fully aware of the implications, she gave her life that others might not die.

The death of the rabies victims, Toos in particular; the increasing number of women, even young girls, who were suffering mental breakdowns; the many ill, whose recoveries were painfully slow because of the shortage or total lack of the proper medication; and the debilitating sameness of each succeeding day lived out within the enclosure of barbed wire and a moat; left us psychologically flagellated automatons. Sensing this, Commander Yamaji was convinced that diversion was needed. All work and no diversion makes for a less effective work force.

A large shed was built to provide the setting for entertainment. Music was allowed. Mrs. Olga de Wit, a white Russian married to a Dutchman, asked for a piano and got it. She was an accomplished concert pianist, known in the Far East for her splendid performances. She organized a Russian choir, writing out the four-part harmony and words from memory. We soon had a large repertoire of Russian songs. Ellie Kucerova, Maus Dol, Tiele Noll, and I were members from Barracks 8. It was marvelously exhilarating and challenging. We sang, in Russian, the happy folk songs of the peasantry as well as some semiclassical selections. The Russian lyrics and melodies were committed to memory, and after much practice we gave periodic concerts.

One evening Olga appeared in a formal—reminiscent of bygone days—on the concert stage. Her dress was so enthusiastically received that the whole camp collaborated on providing formal attire for everyone in the Russian choir. Ladies from Macassar, many of whom had long dresses with them, made them available to those of like size. The ingenuity of the prisoners was amazing; they combined some of the most amazing bits of material. The end results, in the light of the many tiny kerosene lamps, looked not unlike designer gowns. Jet Robyn suggested that I try a dress of hers. It fit, and on the gala evening I stood in the center of the front row, resplendent in an elegant white satin gown. With the sleeves tucked up to make puffed sleeves, the train caught up inside the skirt, and a bright touch of color added at the neckline, you would never have known it was a wedding gown. This was no ridiculous play-acting by a group of POWs traipsing around in formals. It was a deadly serious, well-executed production designed to speak courage and the resolution to live into the hearts of people who for too long had found deprivation, separation, and death their daily companions. Who understands the thrill of seeing the first bright flowers of spring so clearly as one who has just lived through the long, hard winter? Just so, the evenings of song and pageantry were wonderfully enjoyable and therapeutically beneficial to distraught nerves. For an evening at a time we were transported into the realm of the beautiful; we could forget our captivity and our captors.

Two church choirs were organized: one by Mrs. van Veen of Dutch Reformed church members and the other by the Roman Catholic nuns. *Moeder Overster,* the Mother Superior, borrowed my hymnal, and the nuns translated and sang the hymns that had long been a source of inspiration and blessing.

Daisy O'Keefe lived in the Ambon sector of Kampili. Daisy had been a dance teacher in Hong Kong before the war. She instructed some of the teenagers in ballet and other forms of dancing. During their performances, they were accompanied at the piano by Mrs. de Wit.

Books were brought into the camp by the Japanese, and a "library" was organized. The many Dutch books I read were helpful in enlarging my vocabulary.

A bridge competition was organized, and the three best bridge players in camp were Rose David, Lieke Marcar, and Saartje Seth-Paul, all of Barracks 8. None of the diversions interfered with the work program. There were, however, fewer of us to accomplish the same amount of neverending work.

Our evening activities were nevertheless restricted, since everything extra had to be concentrated between the evening meal and lights-out. With the air raids becoming fewer and the special efforts put forth to divert our attention from our circumstances, the weeks passed and Christmas came to Kampili.

We were allowed to hold an afternoon carol-sing and worship service. Carols were sung in Dutch and English; then we were addressed by Father Bell, who was a gifted public speaker. From his first sentence, a hush settled over the audience.

"The story I wish to share with you this afternoon I heard last Christmas for the first time in a gathering similar to this in Pare

Pare. The one who related the story, the Reverend C. Russell Deibler, is spending this Christmas in heaven." Father Bell went on to
tell *The Story of the Other Wiseman* by Henry van Dyke. Through
the graphic descriptions of the narrator, we followed the Other
Wiseman, Artaban, as he sold his richly furnished home in Persia
after having seen the star in its rising. We looked at the three
jewels—a sapphire, a ruby, and a pearl—priceless gems he planned
to carry to the King. We felt his disappointment that none of his
close friends would accompany him on his pilgrimage. We thrilled
as he looked up at the star, then bowing his head, said, "It is the
sign. The King is coming, and I will go to meet him." We moved
westward with Artaban, astride his spirited horse, Baada, along
Mount Orontes's brown slopes, across the plains of the Nisaeans
and the fertile fields of Concabar, over the desolate mountain pass,
through many an ancient city to come at last to the meeting place
of the Tigris and Euphrates Rivers.

Here, nearing his place of rendezvous with the other three
Magi, he came upon a poor Hebrew exile lying across the road,
dying of a deadly fever. We felt Artaban's impatience as the fingers
of the man, holding the hem of his robe, detained him. If he stayed
to give a cup of cold water to the Hebrew, he would miss his
companions, but moved by love, he ministered healing herbs.
Hours later the man recovered. The Jew blessed him in the name
of the God of Abraham, Isaac, and Jacob and told him that the
King he sought would be found in Bethlehem of Judah.

Leaving the last of his food with the man, Artaban hastened on
to find only a parchment under a cairn of stones, stating that the
three Magi could delay no longer—Artaban should follow them
across the desert. We felt his despair in having to return to Babylon
to sell his sapphire and buy a train of camels and provisions for
crossing the desert. We watched his dreary passage over the sea
of sand.

Then from Damascus he moved steadily forward until he arrived at Bethlehem, weary but hopeful of presenting his ruby and
pearl to the King. But he found a city in the grip of fear. The
strangers from the East had left hastily after their visit to Joseph of
Nazareth's lodging. In the dead of the night, Joseph had fled with
the child and his mother to Egypt, some said. Artaban was welcomed and fed in the humble home of a young mother with a

small child. But suddenly the village erupted in a wild confusion of soldiers' clashing swords and the shrieks of mothers whose children were being slaughtered. We thrilled to the compassion that compelled Artaban to part with his ruby as a bribe to a greedy captain, successfully saving the life of the child. The young mother blessed him in the name of the Lord.

We felt the passage of years as the Other Wiseman sought the King in Egypt, taking counsel with a Hebrew rabbi and learning that the Messiah would be a man of sorrow, and that those who seek Him will find Him among the poor and the lowly. He frequented the homes of the lonely and the needy; he fed the hungry, clothed the naked, healed the sick, and comforted the captive.

At the end of thirty-three years, we saw a weary pilgrim, hair whitened, eyes dull, but still seeking the King. He came for the last time to Jerusalem at the season of the Passover and was told that Jesus of Nazareth was to be crucified with two thieves because he had said he was the King of the Jews. Clasping his pearl, he joined the multitude. He would ransom the King with his priceless gem, but at the gate of the city a young girl flung herself at his feet, crying for mercy. She had been seized by rough soldiers intent on selling her into slavery to pay for her dead father's debts. Artaban experienced great conflict of soul, but at last love, the light of the soul, triumphed. He ransomed the girl.

Suddenly the sun was darkened and the earth shook, causing the soldiers to flee in terror. A heavy tile was shaken loose from a nearby house and struck Artaban on the temple. The blow was fatal, and as he was dying, the girl heard a gentle voice speaking to the stricken wiseman. Artaban answered, "Not so, my Lord: For when saw I thee anhungered and fed thee? Or thirsty and gave thee drink? When saw I thee a stranger, and took thee in? Or naked, and clothed thee? When saw I thee sick or in prison, and came unto thee? I looked for thee, but never have seen thy face or ministered to thee, my King."

The same beautiful voice assured him that inasmuch as he had done it unto the least of His brethren, he had done it unto Him, the King. Joy and wonder lighted the face of the Other Wiseman— his journey was at an end; his treasures had been accepted; he had found the King.

Unashamed, I found tears on my cheeks. There were few dry
eyes in the audience. As I left the meeting, I moved toward a new
year inspired by the words of Artaban: "The King is coming; I go
forth to meet Him!"

CHAPTER EIGHT

There was a time when the mention of shock troops sent the adrenaline racing through my veins. However, after our arrival in Kampili, hearing another name totally eclipsed any feeling of fear I had for the shock troops—that of the *Kempeitai!* Even the sight of the sleek, black limousine appropriated by these secret police was ominous and sent a hush over the camp. "For whom are they coming now?" was the question always in our minds when we saw the "death wagon" approaching.

Following closely on the reign of terror perpetrated by the shock troops, the operation of the Kempeitai in Macassar early and officially quelled resistance. Friends, both Dutch and Indonesian, were among the innocents who died cruel deaths. A whispered recounting of those days by one who had had the dubious privilege of observing the *modus operandi* of the secret police confirmed that they were an indomitable foe. They emulated the German Gestapo, specializing in silencing political opposition.

Of the women taken by them from our camp, some returned; others were not seen again. Those who were brought back never spoke openly of their experience. I was to learn why.

An American woman and the Dutch survivors of the bombings on the Island of Ambon had been brought to Kampili. They lived in the Ambon sector. There were but three American women in our sector of the camp, Miss Kemp, Miss Seely, and I. We had all lived and worked in Macassar. We were thankful that Ruth Presswood, an American, was registered as a Canadian on her husband's passport.

One April day in 1944, a chill wind of apprehension swept over the camp as the Kempeitai car was sighted coming from Macassar. Activity automatically ceased, except for an occasional woman leaning toward another to whisper, "Kempeitai." The limousine entered the camp and purred to a stop in front of the headquarters building. We all made a pretense of working, but our gaze was fixed on the office door. Sweet Seventeen came out and hurried to the sewing room. In his wake, Margaret Kemp emerged. Together they walked rapidly toward the Kempeitai car, and Margaret was ushered inside. Sweet Seventeen then broke into a run, headed for the garden area outside the camp. He reappeared followed by Philoma Seely. Before we fully comprehended what was taking place, the two Kempeitai officers emerged from the building and whisked Margaret and Philoma away. We watched in stunned silence.

When life began to pulsate once more through the camp, everyone was troubled by the same questions: "But why them?— What have they done? How long do you think they will hold them?" I dared not voice the questions that were uppermost in my thoughts—"Is it because they're American? Will I be next?"—as though to utter them audibly would make them come true. Fear for the safety of the two women stalked our path from morning till night. Our hearts were bowed in continual intercession for their speedy return to camp.

In an attempt to dissipate some of the tension, the women from Barracks 8 and others of my camp friends planned a surprise party for my birthday. Everyone brought a contribution for the success of the evening. Rice had been pounded for flour to make a very solid cake, which even boasted frosting—raw sugar and black coffee whipped until it had become frothy. It was called *speug,* an indelicate Dutch word meaning "spit," but the flavor was unbelievably delicate! I was overwhelmed, realizing the sacrifice involved and knowing how long it had taken to hoard the sugar and rice needed for this occasion. The families combined their coffee allotments and we enjoyed black coffee with the birthday cake.

The gathering was a wonderful surprise. We tried to make it a gala evening for everyone's sake. Some of the boys pooled their efforts in making a lovely gift for me from a coconut shell. They borrowed (?) the coconut from the commander's kitchen and a saw from the camp carpenter. Removing the top portion of the shell for a lid, they scooped out the inside to make a practical container. Guess what happened to the coconut meat? One of the mothers had fashioned a pompom from colorful bits of wool. I laughed with delight and expressed my gratitude as I examined the highly polished shell with its impertinent monkey-face and pompom stocking cap sitting at a jaunty angle atop its head. I set it on my rack, where I would see it upon awakening. My young friends' ingenuity dispelled the gloom of many an early-morning rising.

Katie, my German friend, had made me a light wool suit from a coat she had ripped apart, turned inside out, then redesigned. One friend knit me a pair of lovely blue anklets. Of course, I had no shoes, but just the same these socks were marvelously bright and greatly appreciated. Someone else fashioned a beautiful red hairband; another lady made one of blue. Handkerchiefs, treasured books, sketches, paintings, mottoes—the list of ingenious gifts went on and on. These expressions of love and affection were a source of joy and a constant reminder that one of God's great gifts is friends.

We closed the evening with folk songs and hymns in Dutch and English. After commending the women and children to God in prayer, we parted with lighter hearts than we had known for a very long time.

Two days after my birthday, on May 12, we had another visit from the Kempeitai. Instantly we were alert, straining to see if Margaret and Philoma were being returned to camp. The secret police stepped out of their car and went into the camp commander's office. The back seat of the car was empty, and I knew in my heart that those men had come for me. I stood frozen, watching; then, as though drawn by a magnet, I started to walk toward the office. Sweet Seventeen met me halfway. My dread premonition was confirmed; I was summoned to Yamaji's office.

They must have heard the thud of my heart. I tried to stand tall and put on a bold front, ignoring their looks and laughing comments punctuated with "America, America." Ceasing to circle me, one began to tap a staccato rhythm on the table top. His black eyes probed my face for the slightest betrayal of a knowledge of Morse code. Then he laid a paper in front of me on which my name had been handwritten—DARLENE DEIBLER. "Are you Darlene Deibler?" he demanded.

"Yes, sir," I responded, "I'm Darlene Deibler, but I didn't write that."

"I know you didn't write it," he spat back. "How well do you know Morse code?" The tapping on the desk continued, insistent, demanding. I replied that I had never studied Morse code and knew nothing about it. The staccato tapping beat on as the Kempeitai officer narrowed his eyes, carefully scrutinizing my face.

"Sir, I know nothing about Morse code," I repeated.

"We'll see what you know about Morse code," he threatened. "Go get another dress and come back here. We're taking you somewhere else to examine you."

In stunned obedience I ran to the barracks, thanking the Lord as I ran that that day I had put on a dress with a full-circle skirt instead of the camp issue of sleeveless blouse and shorts. For some mysterious reason I had felt led to wear that dress that morning. It was the Lord, for I was not given time to change into anything else. I grabbed my Bible from the upper rack, and folded it into a housecoat I had made at Benteng Tinggi. I stepped outside and into the waiting arms of the Catholic Mother Superior. She held me tightly, as though I were a frightened child, and whispered kindly and lovingly, "My dear *Mevrouw* Deibler, we'll be praying for you every day you're gone." Grabbing her hand, I murmured my thanks. Oh, how I needed that! I felt so young, so vulnerable. Then I ran off into the clutches of the Kempeitai.

Crossing the moat, we turned onto the dusty main road to make the fifteen-kilometer drive to the city of Macassar. The car slowed down near a building I recognized immediately as the former native insane asylum, now converted into a prison by the secret police. The car pulled into a circular drive, and as I looked into the first series of barred windows that faced this driveway, I saw someone watching. We stopped and I was ordered out of the car. I stepped out, my eyes riveted to that cell. Margaret Kemp was hanging on the bars, her fists gripping them tightly. Her arms were black and blue. When she saw me, she didn't utter a sound or make a cry. She just shook her head back and forth, back and forth. Looking at Margaret, my heart cried out in protest, "Lord, you took Russell. Must I now go through this?"

Tenderly He answered, *"Whom I love, my child, I discipline."*

Because I had known the quality of His love, my heart bowed in submission, and I whispered, "All right, Lord, all right—just don't leave me." I looked back at Margaret, my first glimpse of the horrors of the Kempeitai, and realized that she was now but skin and bones. Two short weeks ago, when taken from Kampili, she had weighed about 170 pounds.

This former mental institution was a rectangle of buildings containing cells built around an open, central courtyard. I was led into the office, which was in the middle of the two front cell blocks. A young Indonesian woman sat at the desk, and a prison guard stood beside her. He grabbed my Bible and shouted, "You can't have that! All you'll do is sit there and think of that book and what you're reading instead of your evil deeds against the Imperial Japanese Army." He took my wedding and engagement rings and handed them to the girl. They, along with my Bible, were placed in a paper bag.

When taken from Kampili, Margaret hadn't been allowed to carry any extra clothes with her. When the guard left the office, I asked the young woman about the condition of Miss Kemp's clothes. She responded that her dress was in bad shape. I offered my housecoat, saying, "Please, will you give her this?" Rolling it into a ball, she hid it under the desk so the guard wouldn't see it.

The guard returned and, with reluctant feet, I followed him out of the office, through the portico that joined the two initial blocks of cells, and across the courtyard toward another cell block. I could hear a woman crying out in a series of senseless rantings. Immediately I identified the voice as that of Philoma Seely. With chilling clarity, I knew that in two weeks she had become a raving maniac. "Dear God, what have they done to her?"

"Lassie, whatever you do, be a good soldier for Jesus Christ." Dr. Jaffray's last words to me came to mind, and each step became a cry for courage. "O God, whatever You do, make me a good soldier for Jesus Christ."

On the door of the cell before which the guard paused were written in chalk these words, *Orang ini musti mati,* "This person must die." The guard unlocked the door, opened it, and shoved me inside the cell. The door closed upon me, and I dropped to my knees, eyes intent upon the keyhole. When I saw the key make a complete revolution, I knew I was on death row, imprisoned to face trial and the sentence of death.

I listened to the footsteps of the guard recede on the concrete walkway. When he was gone beyond hearing, I sank back onto my heels. My face and hands were wet with cold perspiration; never had I known such terror. Suddenly I found I was singing a song that I had learned as a little girl in Sunday School in Boone, Iowa:

> Fear not, little flock,
> Whatever your lot,
> He *enters all rooms,*
> *The doors being shut.*
> He never forsakes,
> He never is gone,
> *So count on his presence*
> *From darkness 'till dawn.*
>
> Only believe, only believe,
> All things are possible,
> Only believe.

So tenderly my Lord wrapped His strong arms of quietness and calm about me. I knew they could lock me in, but they couldn't lock my wonderful Lord out. Jesus was there in the cell with me.

I inspected my cell. It was about six feet square. The walls were whitewashed plaster and the floor ceramic tile. The barred window had been completely boarded over on the outside to prevent anyone from seeing who was inside. It was the only boarded-over window in the block. A barred transom above the door, which was next to the window, was the only opening in the tiny room. Through this, a vague light shone to illumine my semidarkness. Late that afternoon the guard pushed a tin plate of rice sprinkled with raw sugar into the cell. I wanted to eat; I tried, but couldn't. He grabbed the plate and yelled, "So, you don't like sugar? Well, you'll never get it again."

For seventy-two hours I wasn't allowed to relieve myself. I knew of prisoners who were forced to eat their own excreta when they were unfortunate enough to eliminate on the cell floor! I suffered; I prayed. On the fourth day, the guard finally came to conduct me to the common toilets. I walked out into the sunlit courtyard and through an open prison kitchen area crowded with native male prisoners. Everyone's eyes followed me to the door of one of the toilets behind the cell block. I looked at fecal matter oozing out under the door, and when I pulled the door open, I found the hole in the floor completely hidden by a pile of human excrement nearly calf high. Revolted, I shrank back, whispering, "Lord, how can I step into that? What am I going to do?"

At that moment a young Indonesian prisoner stepped out of the group and said, "Let me wash that down for her." The guard shrugged, moved downwind of the stench, and watched in silence as the young man drew from the prison well the many pails of water needed to flush the hole and then left me in privacy. I returned to my cell braced against every watchful, lustful eye.

Not long after I had been returned to my cell, I heard the guard coming. He unlocked the door and placed a tin plate of rice on the floor. True to his threat, there was no sugar on it. Running my fingers through the rice, I removed the little pebbles, then ate. I knew I must eat to keep up my strength.

Having deftly cleaned off my plate, I was busy licking my fingers when the guard returned with half of a kerosene tin. He dropped it on the floor with a dreadful clang; then, picking up my plate, he left. I examined the tin, wishing they hadn't left the terrible jagged edges when they cut it, but I was so very grateful for it. I wouldn't need to use the filthy common ablution block again.

The dusk had fled before the dense darkness that now crept through the transom. I pushed my back against the wall and waited for sleep to come, but so great was the heaviness of my heart that I cried out with the Psalmist of old, "Why art thou cast down, O my soul? and why art thou disquieted within me?" Why do I feel as if I'm being crushed beneath some awful weight? Here was no one to see or hear my grief. I had no responsibilities, no work roster to prepare or sick to attend—nothing to keep me occupied every waking moment.

My thoughts turned to the Jaffrays, now separated, and to Ruth, who was not strong, missing Ernie so very much. Lilian had become frighteningly frail. None of those missionaries were Americans, as were the three of us imprisoned by the Kempeitai. Not one of us had been involved in political activities. I didn't even know why we were here. What had we ever done that would justify the Japanese's treatment of Margaret and Philoma?

I was hurting; my sorrow seemed unbearable. Then the dam of my tears burst, and I rolled over face down, my bottled-up tears gushing forth like a flood. "Help me, Lord," I cried. "I'm finding it so hard to accept that I'll never see Russell here again. My nights are filled with sweet, happy dreams of him, but I awaken to the reality of his death—already nine months ago."

Sitting up, I tried to stem the flow of my tears. Then I knew my Lord was there, and with my hand in His, we began a walk down the corridor of my memories.

I was reliving a day in 1936 at a district Young People's Rally in Boone, Iowa.

I was the last of the young people to speak. On leaving the platform, I found a seat in the front row. The main speaker was a missionary from Borneo, the Reverend C. Russell Deibler. He held our attention, and many hearts were challenged by his message. As soon as the service concluded, I hurried to meet a friend who had said he would be waiting at the back door.

I was midway down the aisle when a hand grabbed my arm and I heard a lady from the church saying, "Mr. Deibler, here's a young woman I think you'd like to meet." She gave my name and mentioned that I was preparing for missionary service at St. Paul Bible School.

I turned to shake hands and express my appreciation of his message, asking that if he were ever in the St. Paul area, would he plan to speak at our Friday night missionary service? He assured me he would. I thanked him and hurried off to meet my friend.

There was packing to finish and there were goodbyes to be said before I left that week to return to college.

The following spring of 1937 was beautiful. Special meetings were being held throughout the United States and Canada celebrating the Golden Jubilee, the fiftieth year of the Christian and Missionary Alliance. Dr. Robert R. Brown was one of the special speakers to participate in St. Paul and Minneapolis. The missionary from Borneo was also listed. I remembered meeting him in Boone but hadn't thought of him since. They had chosen good speakers, I thought.

The final two weeks before the close of the school year, Mother and Dad made it possible for me to stay in the dormitory so that I might take part in the activities. The dining room was noisy, with girls discussing their graduation dresses and the fellows talking about the sports day. Suddenly I felt that someone was watch-

ing me. I glanced around at several tables before Mr. Deibler caught my eye. He was sitting catercorner across the room at the faculty table, smiling and nodding at me. I tried not to look in his direction after the second time he smiled and nodded. Lunch over, he positioned himself against the wall, waiting.

I don't know why I was annoyed, but turning, I stuck my head into the kitchen pass-through. The cook, seeing me, came over to talk. He was a dear man. Glancing casually over my shoulder, I saw that the dining room had emptied. Thinking the coast was clear, I dashed through the swinging doors to be pulled up short by a firm hand on my arm.

"Just a minute, Miss McIntosh. There's a gentleman here who would like to meet you—Mr. Deibler, Miss McIntosh." With that the dean's wife disappeared; I felt trapped. Why did he want to meet me? I didn't know him, nor he me.

Mr. Deibler immediately reminded me that we had met in Boone. He wanted to tell me that he had stopped in Boone to see my parents "again" before going to Rev. Clarence Anderson's church in Meadow Grove, Nebraska, for their missionary conference. "I had no idea that Mrs. Anderson was your sister until I saw your picture on the piano. I enjoyed my visit with them very much. Your little niece, Coralyn, is a beautiful child, and they all send their love."

Thanking him for visiting my family, I excused myself, saying I had to study. He detained me by asking if I would know who from Boone would be sending him anonymous letters. I thought he was intimating that I might be the culprit. My hackles were up.

"Mr. Deibler, I have never written an anonymous letter in my life, nor do I intend to!" I walked rapidly away down the hall. He certainly had a low opinion of me, I fumed.

"Miss McIntosh, Miss McIntosh, I know you didn't," he said, following me down the hall to apologize. "I just thought you might know who had sent them." I assured him I didn't and left.

The school bus was taking students to Minneapolis that evening for the Golden Jubilee meeting being held in a large theater. The chorale was singing, and space had been reserved for us in the front rows. Mr. Deibler sat on the platform directly in front of me, smiling and nodding. When he spoke, my heart responded to his message. He was a very fluent speaker, and I could understand why he had been detained beyond his furlough time for these special meetings.

Returning to his seat on the platform, he looked at me and motioned with his head and eyes to the exit door near where I sat. I dropped my eyes, pretending not to have noticed.

I didn't want to miss the final speaker, Dr. Robert R. Brown of Omaha. It was through his radio ministry that I, as a child of nine, had first heard that Jesus Christ had died for me. His mention of a church in my hometown, the Midway Tabernacle, drew Mother and me to walk four miles in a blizzard on a dark night in February to find that church and God. We found both.

"My precious Lord, it was that night that You first spoke to me and called me Your child and told me my sins were forgiven. How I love You. We've walked far together."

Dr. R. R. Brown had come the next year, when I was ten years old, for the missionary convention. On the final night, he challenged high schoolers and young adults to give their lives to God for missionary service. My heart was burning. I wished that I were older; I'd volunteer. Feeling a hand on my shoulder, I turned around, but no one was near, and I knew it was my Lord. Quietly I whispered, "What is it, Lord?"

"My child, would you go anywhere for Me, no matter what it cost?" He asked.

The fingers of my soul reached up in wonder to touch Him. "Lord, I'd go anywhere for You, no matter what it cost!" My offering was accepted. God's purpose for my life had dovetailed into the indentations of my longing to do His will. The union was perfect, with no openings to allow the seeping in of uncertainty or frustration as to my future. I had gone to school in St. Paul to prepare for one thing only—missionary service.

Dr. Brown still had a heart that burned for the lost, and many were blessed and challenged by his message. Glancing up after the benediction, I saw Mr. Deibler coming off the platform in my direction. Extricating myself from the crowd, I disappeared through the nearest exit and headed for our bus. Suddenly I felt a tug on my coat, and, turning, saw that it was Mr. Deibler.

"I want to take you to dinner tomorrow evening," he said. I gave a negative response, saying that I was catering a birthday

party the next afternoon and there was no way I could get back here before six-thirty.

He gave me a very determined look, saying, "I'll meet you at the front door of the theater at six-thirty." Not waiting for a reply, he walked away.

"That's what you think! First you insult me, then you want to take me to dinner. Anonymous letter, indeed!" I sputtered to myself, finding a seat in the bus with friends.

They were exclaiming over Mr. Deibler's good looks: "Those dark brown eyes!" "I love that wavy brown hair!" "Did you notice his hands?" Finally one drooled, "I'd give my right leg to go out with him!" Many sighed in agreement.

"Hmm, very interesting! I could go out with him—if I wanted to," I thought. After all, I agreed with them: "He is very good-looking." Probably for no other reason than thinking I had "one-up" on the others, who wanted to go out with him and couldn't, I finished the birthday party in record time and arrived by streetcar in Minneapolis at six-thirty. Mr. Deibler was waiting.

Sitting in a hotel lobby, we talked. "I understand you are interested in the mission field."

I shared with him my call to the mission field and said that it was Dr. R. R. Brown whom God had used to challenge me.

"Do you have any prospects of marriage?" he queried, to which I promptly replied that I did not plan to marry.

"We'll talk about that later," he said as he rose. "Shall we go somewhere for dinner?"

We found a Chinese restaurant nearby. When we were almost finished eating, Mr. Deibler laid down his fork. The silence was portentous. Feeling decidedly uncomfortable, I tossed around in my mind for something flippant with which to break the silence. "A penny for your thoughts."

"Would you really like to know what I'm thinking?" I wasn't sure I did, but he went on to tell me about that Monday at the Young People's Rally in Boone. "You were wearing a lovely brown dress and hat. I could also tell you what you spoke on! I looked up to see your face when you turned to leave the platform and the Lord said, 'That's the girl for you.' I wanted to talk to you; that's why I followed you down the aisle. I'm glad that lady stopped

you to introduce you to me. You certainly were in a great hurry. That afternoon, I vowed I was going to marry you."

Being confused and embarrassed, I protested that he didn't know a thing about me, but I discovered that, thanks to my family and Boone friends, he knew a lot about me.

"I know this is sudden," he continued, "but I want to come back in the summer to see you." Nothing more was said.

The evening service was in progress when we entered the theater and slipped into the empty back row. A few minutes later, Mr. Deibler leaned over and whispered, "Look around." I glanced toward the door to the right, but saw no one. "No, look this way," he said as he laid his hand over mine. I looked up into his eyes and something happened within me. I knew in that instant that I loved him deeply and purely, though I said not a word.

What did I know of this man? I needed to know him. But more importantly, I needed to know God's will concerning any future relationship with him before I admitted any love for him. At the door, I thanked him for dinner and hurried to board the bus.

A special delivery letter from Mr. Deibler was handed me at six the next morning. He was distressed to learn that I was only nineteen and he was twelve years my senior. He feared that the mission board would not accept anyone that young. He was on his way east to talk to Dr. H. M. Shuman, president of the organization. The formal "Miss McIntosh" had become "my darling Darlene," and the letter was signed "Russell."

I knew that if this were the will of God for our lives, I would be accepted and my age would be no barrier. At every opportunity, Russell visited me in Boone, or we met at my relatives, the Ewings, in northern Iowa. There he proposed to me and assured me he had long since had the consent of my parents. Only then did I confess my deep love for him.

Dr. Shuman had lost no time in getting a transcript of my grades and character references from the president, dean, and my teachers in St. Paul. Apparently everything was acceptable, and we were given their blessings.

Russell asked me to meet him in Ohio. The Foreign Board was there to quiz me on theology, doctrine, my spiritual experience, schooling, and my call. When they finished, I was asked to step *outside. Only moments later I was told we had their blessing. I had been accepted at Nyack Missionary Training Institute for my third year of study. They suggested that we marry soon; then Russell could continue to do deputation work while I pursued my studies.*

"And," Dr. Shuman concluded, "Mrs. Shuman and I already have an apartment ready for you in our home. Incidentally, Miss McIntosh, Dr. R. R. Brown came after you went outside, and he said a very good word for you."

Seeing Dr. Brown later in the day, Russell called to him, "R. R., what do you think of my stealing a girl from your district?"

He turned on his lovely Scots brogue and, holding both our hands, he burred, "I couldn't think of a better place to get one!" He, too, gave us his blessing.

Russell had brought his car, because he wanted me to meet his family in Harrison City, Pennsylvania. It was late when we arrived but they were waiting, and I was welcomed by a large, wholesome, loving family. Russell's father, Charles, was a quiet-spoken, white-haired gentleman who lived with Russell's sister Margaret and her husband, Henry Mull. "Marg," as she was called, had been more than a sister to Russell. After their mother's death, Marg had raised him and her own son, Almas Kauffman, whose father had passed away.

I had a happy time meeting the other two sisters, Annie and Ella, and the three brothers, Almas, Jim, and Ralph. It took several days to learn the names of the many nieces and nephews, and who belonged to whom.

Although in spirit I was back in the States, back in years, my physical self was still in the present and in prison. I was still crying, and, having no slip or handkerchief, I had to use the skirt of my dress to sop up my tears. It became soaked, so I stood up to shake it out as I walked back and forth, hoping it would dry. Weary, I sank down onto the floor again, thinking how beautiful our courtship had been, with nothing to mar the memories of it. "How lovely, Lord, it is to make memories without regrets. Help me always to remember this. May I never allow anything in my life that, like the touch of soiled fingers, would leave a stain of regret."

I thought again of Russell's father and the others and remembered their loving goodbyes when I was leaving to return to

Boone. I knew I had found a place in their hearts and was considered "family." Thinking of them, I remembered that they knew nothing of Russell's death. Hurting for them, the tears started again, and I was helpless to staunch them.

Stretching onto the floor, I pressed my face into my wadded-up skirt, lest the night watchman hear my sobs. I continued to remember.

"—and I'll be with you as soon as this convention is over," *Russell called as the train began to move. I realized how blessed* *I was. We had agreed that our wedding should be simple. I could* *make my own dress, veil, and bouquet. Occupied with happy* *thoughts, I was soon back in Boone.*

It was a beautiful wedding. Our church friends asked if they *might decorate. My pastor, Rev. J. H. Woodward, performed the* *ceremony. His wife, Flossie, an accomplished pianist, provided the* *music. Nina Knudsen, a dear friend, agreed to be my bridesmaid,* *and her brother, Earl, was best man. The church was packed, and* *everyone in my family was there.*

More elegant weddings there have been, but none more beau- *tiful. My friends and family helped to make it so, and the weather* *was kind that evening of August 18, 1937.*

We drove back to Pennsylvania that very night. I enjoyed being *with the Deibler family again. They had showers and dinners, that* *I might meet more of Russell's family. We received many beautiful* *wedding gifts, which we took to our apartment in Dr. and Mrs.* *Shuman's home in Nyack-on-the-Hudson.*

Russell was away much of the time, and the Shumans often *remained in New York during board meetings. This left me to* *rattle around by myself in the large, three-storied house.*

When my schedule allowed, I rode with the Shumans to New *York City to meet Russell returning from meetings. The officials* *were busy men but always took time to visit with me. All were* *men with many, many years on the mission field or in the homeland.*

When special prayer requests were received, I saw men who *believed "in everything by prayer and supplication"* *(Philippians 4:6). It was reassuring to a young missionary to* *know that behind me stood men and women of this caliber.*

Before the fall semester finished, Dr. Shuman informed us that *the field had requested an immediate sailing for us. All missionaries* to the Netherlands East Indies were to spend time in the Netherlands in language study. I was assured that my grades were sufficiently high for me to receive my credits. We went home to pack and say goodbye to our families. We were booked to sail from New York on January 27, 1938.

The ship's band was playing "Harbor Lights" as the anchor was hoisted, the lines were cast off, and the tugs nudged the vessel away from the dock. Our friends from headquarters, who had come to say farewell to us, were soon swallowed by the fog. Watching the harbor and the New York skyline recede in the distance, I pondered future harbors and tomorrow's seas. My ignorance of the future held no cause for anxiety, for my spirit witnessed within me that God was and would be in control. With exhilarating joy, I turned my face into the wind blowing off the Atlantic.

Notwithstanding the rough seas, it was a restful voyage and a precious time for Russell and me to be together. We prayed much about our future ministry, and constantly our thoughts during prayer and discussions turned to New Guinea.

Arriving in Holland, we wrote Dr. R. A. Jaffray concerning our desire to be appointed to New Guinea. A letter from Dr. Jaffray crossed ours in the mail, asking if we would be willing to accept an appointment to New Guinea. We never questioned again God's leading us to New Guinea.

We progressed well in our study of the Dutch language. I loved the country and its people. We made many friends and carried away wonderful memories. Traveling via London and Paris, we proceeded to Marseilles to board the ship for the Indies.

Back in the present, dawn was interjecting itself into the cell. Brushing tears away with my hands, I wiped them onto my arms. I was determined, God helping me, that the Japanese would never see or hear me cry. I knew that I must rise and try to get my dress dry. Moving slowly back and forth, I waved my skirt in the air. I dared not remove my dress as I had no slip, nor did I know when my cell door would open suddenly. Miraculously, my skirt was dry when I stood to receive my food. I could hear the guard coming.

After I had eaten and my plate had been collected, I slumped against the wall. When I came into the Kempeitai prison, I knew I had dysentery. Now the debilitating effect of dysentery was already draining my strength. I wished for a cover for the kerosene tin.

More and more flies were coming in through the transom. I had contracted it during the outbreak at Kampili but had kept going because of the work quotas.

Since there was no cover for the kerosene tin, a squadron of large blue-bottle flies were attracted by the dysentery. When my plate of plain rice was brought, the flies filed squatters' rights to it before the tin plate had stopped spinning across the floor. Because it appeared that I was to be fed only once a day and the portions were small—less than a cupful—I decided not to share. I fought off those dirty flies—how I hated them—and hid the rice under my skirt. I'm very fond of rice, and the fact that they gave me no spoon or fork didn't disturb me. I had learned to eat rice à la Indonesian—with my fingers. Reaching in under my skirt, I rolled some rice into a ball, waited until the flies were otherwise occupied, then popped it into my mouth. It became a daily contest between us, but I was usually able to outwit them.

Weary, I lay down and began to count the months since our marriage of just over six years had been terminated by sickness and cruelty. I willingly accepted the year and a half we were separated, both in the States and while Russell was in New Guinea establishing our mission base. "But Lord, being robbed by the Japanese of the last year and a half of our marriage—that is very, very bitter," I sobbed. "Please, Lord, take this bitterness from me. I have so much of good and so many beautiful memories that nothing and no one can ever take from me."

The Man of Sorrows, the One acquainted with grief, enfolded me in His arms. As He poured in the comfort of the Holy Spirit, the anger and the bitterness were released and I slept.

In the dark hours of the night, Russell came to me in a dream. We were aboard ship, sitting on deck. We talked as he held me close. Taking up a basket that was there beside him, he took out a jar of sun cream. "This, dear, you must keep with you. Your skin is so sensitive to the sun. Now I must leave you." Rising, he committed me to the One Who would never leave or forsake me, and he was gone. I never dreamed of him again.

Of course, when the guard had the slop bucket emptied the following day, he knew I had dysentery. The treatment was a glass of Epsom salts. How I hated that nasty solution! Further, porridge was substituted for the rice I had been receiving. This gruel was dirty rice boiled in water and not flavored with salt, sugar, or milk. The first day I received it, I saw what appeared to be shredded coconut floating around on the surface. "O joy, I love fresh shredded coconut," I thought. Closer inspection caused my joy to be short-lived. Those were not shreds of coconut; they were worms. There was a sprinkling of small stones and chaff as well. Carefully pushing those choice ingredients up onto the side, I lifted the enamel soup plate and let the porridge drain between my hands into my mouth. Knowing that I could have been left without any food, I gratefully thanked the Lord, worms notwithstanding.

The flie were maddening. I couldn't manage the porridge under my skirt as I had the rice. By the time I had the worms on the perimeter of the plate, the flies were there feeding on them. I needed both hands for funneling the porridge into my mouth, so I couldn't shoo them. In but a day or two, I came to the wise decision that if the flies could eat the worms, so could I. I shooed flies with one hand and removed only the stones with the other. With a mighty flourish of my hands, I brushed my tormenters aside and as quickly as possible devoured the porridge, worms, and whatever. Then I licked the plate and my hands clean in order not to miss anything.

Nights on the coast are cool. With nothing but the cotton dress with the full-circle skirt to cover me, and lacking a mosquito net, I was generally miserable for sleeping.

Early in the first week, I passed the time killing mosquitoes. I was tortured by hordes of them at night. They clung to the wall, too full of my good red blood to do anything else. Wherever I made a kill, a splattered blood mark was left. One day when the guard brought my rice, he suddenly noticed the splotches decorating the wall—I had had an unusually heavy kill that morning. Infuriated, he ordered me to remove every mark, and to have it done before he returned. I swallowed my porridge, not even stopping to look for stones, then gave my plate and hands a lick and a promise before setting to work. I ate first, because when he returned it would be for my plate. If I were to be beaten, it might be easier to take on a partially filled stomach. Fear is a great persuader: laboriously and feverishly, I scraped away all the dried spots of blood. When he did return for my plate, the splotches were gone—and so were my fingernails. I had worn them down to the quick. Having learned my lesson, from that day on I killed my small tormentors by crushing them between my hands. I became quite adept at it, and my new technique kept the blood off the wall.

The guard brought in two planks nailed to cross-beams to serve as a bed. The mosquitoes were especially bad the first night on my new bed. They even seemed to get beneath the bed to attack me. Whoever made the makeshift bed had left a one-inch gap between the planks, but uncomfortable though it was, I was grateful. The planks kept me up off the cold ceramic tiles at night.

Noticing bites all over my body the next morning, I slid over to the boards to examine them. Just as I had suspected, they were crawling with bedbugs. I dislodged the vermin from the grooves with a hairpin, then crushed them as they tried to make their escape across the floor. When I finished, I scraped together whatever was left of their bodies, threw them in my tin, then spit on the floor and rubbed it with my hand until all trace of my kill had disappeared.

I was grateful for the ceramic-tiled floor. I couldn't have killed the bedbugs on a dirt floor with as much ease, nor could I have kept a dirt floor clean. How dirty my dress would have become, sitting on a dirt floor! Also, the tile was refreshingly cool during the heat of the day. There was a space under the door through which I could blow to the outside the dust I had scraped together. When all the killing and cleaning were finished, I spit-bathed my hands—and with a measure of finesse, rubbing and rolling the dirt off into the tin. I had learned a valuable lesson from my New Guinea friends. If I hadn't known to do this, my plight would have been compounded, for I never received soap or so much as a drop of water for washing.

The days were spent in contemplation of the goodness and presence of my God and in listening to the noises of the other prisoners. I listened to those who were allowed to go into the courtyard as they chatted in the sunshine. I listened to the sound of footsteps on the concrete walkway and learned to distinguish the padded sounds of the guard's tennis shoes from the hard clicks of the leather-heeled boots of the officers. I listened to the guard coming to get others on death row. I listened to the torture, as prisoners screamed, pleading for mercy. I fell to my knees, covered my ears with my hands against this horror, and cried out to God to undertake in His mercy and spare their lives. I heard their limp bodies dropped like wet sacks to the floor. The sound of silence then meant that the prisoners were unconscious. I listened when prisoners were dragged along the walkway past my door, listened to the cell doors being opened and the victims dragged inside and left to regain consciousness in the solitude of their cells. I listened to their sobbing and moaning—signals of their return to consciousness.

According to radio reports from countries that had been invaded before the Netherlands East Indies fell, we had learned that the Kempeitai had many forms of torture used for extracting information. None held greater horror than the water treatment. The hapless one who was to receive the water treatment was pegged spread-eagle on the ground, face up. A rubber tube was forced down the prisoner's throat and into his stomach. Water was then poured through this tube into the stomach until it became horribly distended. The secret police then jumped on the bloated abdomen or rammed the stomach with a cane until the desired information was forthcoming or the victim fainted. Irreparable damage and internal hemorrhaging resulted from this form of torture.

One of our young Dyak students was tortured at the Kempeitai headquarters prison. He said he was forced to hold high above his head a five-gallon tin of water, and while all the muscles were stretched taut, they rammed him in the stomach with their canes. He was completely innocent of any subversive activities, but by then the damage was done. He suffered intense pain, but he was one of those God could trust with pain. He could still praise his glorious Lord and forgive those who caused his death—for he died shortly thereafter of his internal injuries.

If prisoners were not returned to their cells immediately, I had to will myself not to think about the water treatment—I could be the next victim. At such a time as this, I took refuge in God's Word, especially Psalms 27 and 91.

The most malignant sound to my soul was the ranting of Miss Seely. To hear her this way, realizing that she had probably been mistreated to force a confession about my alleged crimes, was a burden most unbearable. Day and night she called out phrases in Indonesian. I think she was trying to sing, but because she was tone-deaf, there was no melody, just an awful, raucous sound coming from her cell. For four days I was left alone to absorb this cacophony of horrors—a part of the softening-up process.

The next day the Japanese guard led me to the interrogation room for the first time. The two men whom I had met in Kampili were waiting for me. The guard was the number three man of the examining team. I mentally labeled the tall, cruel man "the Brain," for he was the brains of the team. He formed the verbal bullets, and "the Interrogator" fired them. The Interrogator was small and slim, with a shock of curly black hair and a large, forceful hand. The Brain didn't look Japanese; not only his height, but his features were very different. His eyes were piercing black pools of pure, unadulterated hatred.

The Brain and the Interrogator sat at tables, and I was placed on a chair between them. With pen and paper before him, the Brain always positioned himself on my left, just out of my line of vision. He could then observe my every expression and response to their questions. The Interrogator sat directly in front of me and within arm's reach. The Brain never addressed me directly. His questions were fed into the Interrogator in Japanese, who in turn fired them at me in Indonesian. My responses were in Indonesian. Because of this procedure, I assumed that the Brain didn't know Indonesian.

If an answer didn't please the Brain, he would snap an order to the Interrogator, who struck me a swift karate chop with the cutting edge of his hand. He was skilled in striking with great force and precision that very tender spot close to the shoulder at the base of the neck. A well-placed blow was excruciatingly painful and paralyzing. Other times the Interrogator used his middle finger. Flicking it from behind his thumb, he thumped me repeatedly between the eyes until I thought my nerves would explode. My forehead became bruised and swollen. Often I felt I couldn't endure to be touched again, but the Lord comforted and strengthened me. I remembered Dr. Jaffray's injunction: "Lassie, whatever you do, be a good soldier for Jesus Christ." I prayed that God would help me.

The hearings—for my first interrogation was only one of many—lasted an hour, two hours, sometimes longer. All the questions were relative to my alleged espionage activities. What contact had

I had with the American military personnel? What knowledge did I have of Morse code? (Because they reverted to this question so often, I began to think there was a radio somewhere that they were trying to find.) Who were my superiors? Where had I been groomed to become a spy? Where did I get my radio? What did I do with it when I was moved to Kampili? Who were my contacts? If I didn't have a radio, how was I getting information out of the camp?

On and on they went, infinitely and with enormous repetition. I prayed constantly that the Lord would help me to answer carefully and briefly. They tried to brainwash me into thinking I had been used by the Americans to spy for them and that the Americans had betrayed me to the Japanese. They would be lenient if I confessed, they said, but I knew that in war time spies were beheaded. It was a psychological pummeling that left me "bloodied but unbowed." I gritted my teeth and vowed, God helping me, that they would never break me. This experience was to be repeated every few days for the next month and a half. I never knew when to expect them to call me to the hearing room again.

I never shed a tear before them during the hearings. But when the guard had returned me to my cell, and the sound of his footsteps had vanished—when I was certain that no one could hear me—I wept buckets of tears. In desperation I poured out my heart to the Lord. "O Lord, I just can't go through another one. I *can't,* Lord, I just can't. Please, no more, Lord."

When there were no more tears to cry, I would hear Him whisper, "But my child, my grace is sufficient for thee. Not *was* nor *shall be,* but it *is* sufficient." Oh, the eternal, ever-present, undiminished supply of God's glorious grace!

Just two weeks before I was brought to this prison, the Lord had laid it on my heart to memorize a poem by Annie Johnson Flint. Now I knew why. After drying the tears from my face and mopping the tears from the floor with my skirt, I would sit up and sing:

> He giveth more grace when the burdens grow greater.
> He sendeth more strength when the labors increase.
> To added affliction, He addeth His mercy,
> To multiplied trials, His multiplied peace.

When we have exhausted our store of endurance,
And our strength has failed ere the day is half done,
When we reach the end of our hoarded resources,
The Father's full giving is only begun.

His love has no limit, His grace has no measure.
His power has no boundary known unto men.
For out of His infinite riches in Jesus,
He giveth and giveth and giveth again.

Strength came, and I knew I could go through another interrogation, and another, and another. I was physically weak, and desperately frightened, but God gave me the courage to deport myself like a good soldier for my Lord before those cruel men.

One day I glanced longingly at the transom above the door. *If only I could get some air.* I pulled myself hand-over-hand up the bars of the window until I could lunge and catch the bars over the open transom. I managed to balance, one foot on the doorknob, one on the window sill. Hanging from the transom bars and balanced thus precariously, I could look out on the open courtyard. Because of the roof overhang jutting out over the walkway in front of this block of cells, I could see out but nobody could see me. The cooling breeze blowing off the ocean was refreshing, and the sight of something other than my four walls was exhilarating. "I must do this again," I promised myself, "but now I need to conserve my strength."

The mosquitoes had their revenge. I fell ill with cerebral malaria, a type of malaria that is often fatal. I burned and chilled. When the chills came, I wrapped my arms and the skirt of my dress around me to try to still the violent shaking; it was almost a relief to be awash in the waves of the burning fever that followed. To the Epsom salts administered for the dysentery, the guard added a glass of water in which powdered quinine had been dissolved. Swallowing those wretched, bitter solutions was torture in itself. The rank taste of quinine curdled in my mouth all day long. I longed for a swallow or two of plain water, but that was denied me. I began to suffer from the debilitating symptoms of beriberi, a deficiency disease caused by lack of vitamin B_1 in the diet. My legs were becoming badly swollen; the tissues were filled with fluid. Daily I checked them, pressing my thumb into my leg to feel for the shinbone. I counted to see how long before the deep indentation disappeared and the tissues filled with fluid again. It was laughable—the contrast between my large, shiny legs and the emaciated condition of the rest of my body.

Scratches underneath the window ledge, hidden from the guard's eyes, helped me keep track of the passing days, and with each new one I could feel my strength waning due to the beriberi. Would I also suffer paralysis and anemia? When I arose in the morning, I would walk around my tiny cell one way until I got dizzy and then around in the opposite direction, in a determined effort to keep up my physical strength. Much time was passed repeating Scripture. Starting with *A,* I would repeat a verse that began with that letter, then on through the rest of the alphabet. I discovered that most of the songs we had sung when I was a little girl were still hidden in my heart, though I hadn't consciously memorized many of them.

As a child and young person, I had had a driving compulsion to memorize the written Word. In the cell I was grateful now for those days in Vacation Bible School, when I had memorized many single verses, complete chapters, and Psalms, as well as whole books of the Bible. In the years that followed, I reviewed the Scriptures often. The Lord fed me with the Living Bread that had been stored against the day when fresh supply was cut off by the loss of my Bible. He brought daily comfort and encouragement— yes, and joy—to my heart through the knowledge of the Word.

Paul, the apostle, wrote that it was through the comfort of the Scriptures that he had hope and steadfastness of heart to believe God. I had never needed the Scriptures more than in these months on death row, but since so much of His Word was there in my heart, it was not the punishment the Kempeitai had anticipated when they took my Bible.

Late one afternoon of a day when I had not been interrogated, I heard laughter and animated conversation filtering in through the bars of the transom. I thought, "Hey, I wonder if that isn't the girl who was in the office the day I was brought from Kampili. She was kind enough to deliver the housecoat to Margaret. She sounds happy. Now who around here could make anyone happy?" Curiosity impelled me off the floor. I felt delight that someone was happy and longed to witness it, even if I couldn't share it. I climbed to grab the bars and pulled myself into position. There on the ledge was a knife! By sheer effort of will I hung on, my scalp crawling. "Dear God, where did that come from? Who put it there? When? Could it have been while I was in the hearing room?"

I slid to the floor, thoughts of happy people completely blocked out. Who? When? Why? Were my captors expecting me to commit suicide? Then they wouldn't have to continue the interrogations— all those ridiculous trumped-up charges. Or was that knife put there as evidence that I was in contact with someone on the outside who had brought it to me, and I had hidden it there? I tried to remember all I had heard and read of Kempeitai dealings with prisoners, and a score of other possibilities paraded down the halls of my mind to be considered and explored. I planned to attack the guard and attempt an escape—oh, the possibilities were legion.

I stretched out flat on my back and breathed deeply, trying to still the nauseous churning of my stomach. "Lord," I prayed, "I need counsel. Should I hide the knife? If so, where?" No, they would be sure to search the cell, and there were no hiding places. If they found the knife, my problems would be compounded. "Should I wait until dark and throw it out into the courtyard?" No good. They would be sure to find my fingerprints on it. My finger-prints were all over that door and window. With extreme care, using the skirt of that wonderful all-purpose dress, I cleaned the bars, the transom, the doorjamb, and the window ledge. Then on my knees, with my face to the floor, I explained the whole hopeless situation to the Lord. With the telling, quietness invaded my spirit and I began to worship the God of Abraham, Isaac, and Jacob, the God of Elijah and Daniel, the God of miracles. "Lord, if You could open the Red Sea to deliver Your people from Egyptian tyranny, and if You could send Your angel to shut the mouths of lions that they might not kill Daniel—then, Lord, it is nothing to You to remove that knife. Thank You, Father."

For three days I never left the cell. No one came or went without my notice. None could have reached the knife without a ladder or without my hearing. Late in the afternoon of the third day, I crawled up to find an empty ledge. The knife was gone! "— and Father, erase all memory of it from the mind of whoever put the knife there." He did just that, bless His holy name. I didn't even try to figure it out. I just knew it was the Lord. I let God be God, and truly believed that with Him all things were still possible.

An older Indonesian night watchman was added to the night guard. The office girl had been chatting with him the afternoon I found the knife. Time would tell what kind of a man he was, and I began to pray for him.

A few days later, I heard the guard unlocking doors adjacent to mine. Who or how many were being brought in, I couldn't tell. After the guard left, I heard the new prisoners call to one another, and I was certain of the identity of two of them: Saartje Seth-Paul and her sister, Bea, from Barracks 8. I didn't recognize the voice of a third person. Cautiously I took up a listening post at the transom. I was afraid to make my presence known, lest the other woman be an informer. However, when I heard Saartje call "Joop," I knew it was the Dutch nurse who, being taken through Benteng Tinggi to Malino, had brought Ernie's pen and letter to Ruth. She had told me then that she had seen Russell in Macassar and he had sent his love to me. I was in a fever of excitement; here were friends I could talk to without fear of incriminating myself. But my pleasure was dissipated by thoughts of Oma de Graaf and of Saartje's children—how terribly distressing for the family!

It emerged that the Japanese were trying to uncover a large sum of money that the Seth-Pauls were reported, by an informer, to have hidden. Hidden or not hidden, I never asked. With vicious applications of the cane, the Japanese were trying to extract the whereabouts of this purported cache. The bruises on Saartje's and Bea's arms and faces were plainly visible. They stood under my transom and showed me the shocking contusions on their backs and legs. Horrible, cruel, vicious, unprincipled, immoral, inhuman beasts—my repertoire of appropriate adjectives to describe the Kempeitai grew mightily in those days.

I was very proud of Saartje and Bea when they emerged from the interrogation sessions, heads held high—bloodied, too, but unbowed. One afternoon the guard allowed me to take a dipper bath when the three Dutch women were taken for theirs. I had almost forgotten the luxury of a water bath. During the time of their stay on death row, I wasn't taken to the hearing room. I don't know of what Joop was accused, but before the week was out all three women were returned to Kampili.

The debilitating and embarrassing effects of the dysentery were a great burden to me. The flavor of the quinine and Epsom salts had not improved. "Lord, I'm being constantly reinfected by the flies, so if it please You, heal me." Faith welled up in me to believe that the Lord had seen my distress and heard my prayer. When the guard brought my daily portions of quinine and Epsom salts that day, I refused to drink them. Instead, I witnessed to him about the peace God gives to those who believe in Him. I also explained that I was trusting Him to heal me. He insisted that I drink the medicine, so I took the glasses but never drank from them. As the day went on, I was convinced that the dysentery had been overcome by the healing power of God. The Great Physician passed by, and by faith I reached out to touch the hem of His garment.

When the guard returned to collect the glasses, he chided me for not having drunk the medicines. I replied, "I told you, I don't need them anymore. God has healed me." He took a stool specimen to the prison clinic. The report came back: not a trace of dysentery! It wasn't until later I realized that I had asked the Lord to heal me only of dysentery, but He had also healed me of malaria and the beriberi. No more fevers and chills, and the swelling began to go down in my legs. It was such a glorious release to be delivered from the physical symptoms of those diseases.

One time, after I had been in the Kempeitai prison for several weeks, facing my interrogators every few days, I walked into the hearing room and looked into the eyes of a Chinese man who appeared surprisingly familiar. Recognition came: it was the man who had walked out of the jungle the day Margaret Jaffray and I had strolled down that peninsula of land to check on Benteng Tinggi. I remembered my fleeting impression—*you'll see this man again.* Sure enough, there he was. He now accused me of spying for the American cause. He witnessed that he had seen me in the jungle. I had been checking on the movement of troops. He had seen the radio in my hands. I was astounded! My hands had been empty on that day—I had carried nothing.

I attempted to explain why I had made that ill-fated walk. I was simply curious—curious to see if the houses at Benteng Tinggi were still standing. "Curious" had been dropped from the Japanese vocabulary. There had to be a subversive reason for my unexplainable behavior. Holding a watch up to my ear to test my hearing, the Interrogator leveled charges more ludicrous but at the same time more serious.

From this point forward, my interrogations in the hearing room centered on the spoken witness of that man, who I presumed had received something for his false testimony. "Why did you want to know if we had soldiers over there?" "Why were you watching?" "Why were you down there away from the other women?" "Where is the radio now?" "Have you turned it over to an underground agent?" "If you didn't have a radio, who is your contact?" "How did you get information concerning troop movements to the Allies?" Check and recheck—had I told the same story as before? None of my answers was satisfactory. Finally, around the sixth week, I was informed that the Kempeitai had positively established the fact of my involvement in espionage. They refused to believe anything I had said.

The Interrogator looked at me and asked, "You know the penalty, don't you, for espionage work in war time?"

I knew: I was condemned, without formal trial, to be beheaded as an American spy.

That afternoon the tears flowed more copiously than ever before. I felt the pressure of His hand upon the hurt; the anguish receded before the calm, reassuring quietness of knowing that His grace would sustain.

After this session, I was never again taken to the interrogation room. I lifted my heart in praise and thanksgiving to the God Who enabled me to endure. He had given me the strength to be a good soldier for Jesus Christ. What was to be in the future I left completely in His hands. Even my cell took on the beauty of a place of refuge—I didn't have to go back to the hearing room again.

About a week later, at the time of day when the guard herded the women from the front cell block into the courtyard, I decided to check on Margaret Kemp. There she was in my housecoat. I was so pleased that she had it and wished I dared call to her. With her in the graveled courtyard were several native women prisoners. They had been jailed for minor misdemeanors and were allowed to take air and exercise afternoons in the courtyard, whenever it pleased the officer in charge.

The actions of one woman in particular fascinated me. Every time the sentry on duty turned his back to her and marched to the other end of the courtyard, she inched over toward a fence covered with Honolulu Creeper. When the guard clicked his heels, turned about, and began to stroll in her direction, she stopped. There he went, and there she went. "Aha, intrigue. She's going to make contact with someone who's hidden in those vines. Isn't this exciting! Oh, do be careful. With no books to read, I'll watch the drama taking place here before my very eyes!" I empathized with her. I wanted her to succeed, and not to be caught. Finally, reaching the vine-covered fence, the woman stood very still. The guard clicked his heels and went off again. At that moment, I saw a hand shoot through the tangle of vine. It held a big bunch of bananas. Quickly she grabbed the bananas, slipped them into the folds of her sarong, and strolled nonchalantly back to join the other women. Nobody knew she had those bananas. But I did—*bananas!*

I dropped to the floor of my cell. Exhausted from my efforts, I shook all over. Worse still, I began to crave bananas. Everything in me wanted one. I could see them; I could smell them; I could taste them. I got down on my knees and said, "Lord, I'm not asking You for a whole bunch like that woman has. I just want one banana." I looked up and pleaded, "Lord, just *one* banana."

Then I began to rationalize—how could God possibly get a banana to me through these prison walls? I would never ask the guard. If he helped me and was discovered, it would mean reprisals. I would certainly never ask a favor of the Interrogator or the Brain. There was more chance of the moon falling out of the sky than of one of them bringing me a banana. Then I ran out of people. These three were the only ones. Of course, there was the old Indonesian night watchman. "Don't let it even enter his thinking to bring me a banana. He'd be shot if caught."

I bowed my head again and prayed, "Lord, there's no one here who could get a banana to me. There's no way for You to do it. Please don't think I'm not thankful for the rice porridge. It's just that—well, those bananas looked so delicious!"

What I needed to do was link my impotence to God's omnipotence, but I couldn't see how God could get a banana to me through those prison walls, even after the knife episode and my healing.

When the Japanese officers from the ships docked in Macassar Harbor visited the prison, great hardships were inflicted upon the prisoners. We were laughed at, scorned, and insulted. When our cells were opened, we were expected to bow low at a perfect ninety-degree angle. If we didn't perform to their satisfaction, we were struck across the back with a cane. These were humiliating and desperate experiences.

The morning after the banana drama, I heard the click of officers' leather heels on the concrete walkway. The thought of getting to my feet and having to execute a bow was onerous, to say the least. My weight had dropped during those months in the converted insane asylum, until now I was skin drawn over bones. One nice thing about my streamlined proportions was that the thinner I got, the longer my dress became, so I had more covering at night. I stretched out my hands often and laughed at my bird's claws. The meager daily meals were not designed for putting on weight. I had been healed, but I needed food for strength. I wondered if I could manage to get to my feet and remain upright, but I was determined that when that door opened, they would find me on my feet.

The officers were almost at the door. I reached up, grabbed the window ledge, and pulled myself upright. "Now, Lord," I prayed, "officers are coming. Give me strength to make a proper bow." I heard the guard slip a key into the door, but he had the wrong one and ran back to the office to get the right key. I dropped to the floor to rest, then came to my feet again when I heard his tennis shoe–shod feet moving quickly down the walkway. My legs were trembling, and I clutched the bars of the window to steady myself. "Lord, please help me to bow correctly."

Finally the door opened, and I looked into the smiling face of Mr. Yamaji, the Kampili camp commander. This was early July, and it had been so long since I had seen a smiling or a familiar face. I clapped my hands and exclaimed, *Tuan Yamaji, seperti lihat sobat jang lama,* "Mr. Yamaji, it's just like seeing an old friend!"

Tears filled his eyes. He didn't say a word but turned and walked out into the courtyard and began to talk with the two officers who had conducted my interrogations. At roll call in Kampili,

I had had to give certain commands in Japanese, but I had made a deliberate effort to learn as little of the Japanese language as possible. It was better not to know it. I couldn't understand what Yamaji was saying—but he spoke with them for a long time. What had happened to the hauteur and belligerence with which those two always conducted themselves toward me? I could see their heads hanging lower and lower. Perhaps he spoke to them of my work as a missionary, or maybe he shared with them concerning that afternoon in his office after I had learned of Russell's death, when I spoke of Christ, my Savior, Who gives us love for others— even for our enemies, those who use us badly.

Finally Mr. Yamaji came back to my cell. "You're very ill, aren't you?" he asked sympathetically.

"Yes, sir, Mr. Yamaji, I am."

"I'm going back to the camp now. Have you any word for the women?"

The Lord gave me confidence to answer, "Yes, sir, when you go back, please tell them for me that I'm all right. I'm still trusting the Lord. They'll understand what I mean, and I believe you do."

"All right," he replied; then, turning on his heels, he left.

When Mr. Yamaji and the Kempeitai officers had gone and the guard had closed the door, it hit me—*I didn't bow to those men!* "Oh Lord," I cried, "why didn't You help me remember? They'll come back and beat me. Lord, please, not back to the hearing room again. Not now, Lord. I can't; I just can't."

I heard the guard coming back and knew he was coming for me. Struggling to my feet, I stood ready to go. He opened the door, walked in, and with a sweeping gesture laid at my feet— *bananas!* "They're yours," he said, "and they're all from Mr. Yamaji." I sat down in stunned silence and counted them. *There were ninety-two bananas!*

In all my spiritual experience, I've never known such shame before my Lord. I pushed the bananas into a corner and wept before Him. "Lord, forgive me; I'm so ashamed. I couldn't trust You enough to get even one banana for me. Just look at them— there are almost a hundred."

In the quiet of the shadowed cell, He answered back within my heart: *"That's what I delight to do, the exceeding abundant above anything you ask or think."* I knew in those moments that nothing is impossible to my God.

After God assured me it was His delight to send me those bananas, my heart was salved, and it took all the character I possessed not to eat all ninety-two in one sitting. After months of meager rations of rice porridge, I knew that to gorge could make me deathly ill, so I portioned out so many bananas per day, saving the greener ones for last. This was God's provision, and strength began to flow into my body. "Thou preparest a table˜before me in the presence of mine enemies" (Psalm 23:5).

Late in the afternoon of the day following Yamaji's visit, I heard the Indonesian night watchman pause outside my door. He called softly, *Njonja.*

"Yes, sir?" I jumped up and put my ear to the door.

Njonja suka pisang gorengkah? "Do you like fried bananas?"

"Oh, yes, I like anything to eat!" I heard him walk away and knew he wasn't risking opening the door. Trembling with excitement, I climbed to the transom above the door and saw him returning. The overhang of the roof concealed from the view of others the fried banana tied in a corn husk dangling gaily from the bayonet atop his gun. Imperceptibly he shortened his steps as he passed my cell to allow me to reach out through the transom and grab the gift. Murmuring my sincere thanks, I jumped to the floor and promptly dispatched it with relish. Did Mr. Yamaji's act of kindness embolden the night watchman, or was it pity for me that prompted the gift? Perhaps both.

The next day I put my finger in the porridge to remove a black speck and uncovered a fish eye leering at me—in fact, ten little fish eyes, ten little minnows all in a row. It was very exciting. Had the cook hidden them in the porridge for me, or had someone else? I never learned, but it "warmed the cockles of my heart," as Dr. Jaffray would say, to know that someone out there cared.

A few days later, after the guard had removed my plate and as the sun was withdrawing from the cell through the transom, I sat languidly contemplating my situation. Why had they not already had the execution? Why had they not returned Margaret and Philoma to Kampili? Did the Kempeitai not want the people of Kampili to see what they had done to them? At least I knew that Philoma was still here. I decided that I must climb up to the transom the next day to see if they brought Margaret out with the other women; if not, she might already have been returned. What was the news of the war? Had the U.S. Navy really been sunk, as the occasional visiting Japanese admirals had informed me? But then, they had *"just completely sunk* the U.S. Navy" on three different occasions in the past two years. I chuckled to myself: good old U.S.A., she's making her naval vessels of cork these days—they're always bobbing up to be sunk again, and yet again, to have more Japanese ammunition expended on them. "Watch for them; they'll surface," I said to myself.

Gradually I drifted into the spiritually unprofitable game of "suppose"! *Suppose* the Japanese do win the war, what then? That I'll never believe! But none of us believed Hong Kong or Singapore would fall. They fell, and in just a few days. Kamikaze pilots are formidable foes. *Suppose* my brothers Donald and Ray are here somewhere in the South Pacific, facing the kind of enemy who would use flame throwers on nurses huddled in a cave, the kind of people who would do what these men have done to Philoma and Margaret, innocent missionaries. I was innocent, too, but then they did have a witness. It could be that he had been caught hiding out in the mountains, so maybe he was fighting for his life, too. We had heard that some of the Macassar Chinese were ill-treated. I felt compassion for my accuser well up within me. "You poor Chinese man, to lie in your testimony against me." I prayed for him. *Suppose* Don and Ray are killed, what of their families? What of Mother and Dad? *Suppose* none of us makes it home?

There is nothing that will plunge a person into despair more quickly than to *suppose* what could happen. This was another example of the worries of tomorrow that never come, robbing us of the joys of today. Poignant sadness, overwhelming me for the hurt of others, released the tears for my own widowhood. I was alone and I had time to weep, but with the tears came healing. In my moment of terrible aloneness and sorrow for a world of people so devastated by war, I heard someone with a beautiful, clear voice singing "Precious Name, Oh, How Sweet" outside my cell, but he was singing in Indonesian: "Precious is Your name, a shelter that is secure!" My heart burst with bright hope! The "time to weep" was past; it was a "time to laugh."

"O Lord," I cried, "forgive me. It isn't a game of 'suppose.' I live in the sure knowledge that "the name of the Lord is a strong tower: the righteous runneth into it, and is safe" (Proverbs 18:10). The name of Jesus, Your precious name, is my strong tower of defense against the enemy of despair. It is my shelter that is secure; I enter in and am safe."

But who was the singer? How could he know I needed that song at that moment? Of course he couldn't know, but he loved God, that is sure. I had to see him.

I scrambled up to the transom. My eyes probed the late afternoon light—no one by my door, no one in the courtyard other than the guard and night watchman. They were talking, and I knew they were totally unaware of the singing! Listening to this hymn of hope and assurance coming from I knew not where, great awe filled my heart. Quietly I slipped to the floor and bathed my soul in the presence of my God.*

Sometime after midnight, I awoke with a start and was instantly on my feet by the door. Someone wearing tennis shoes was moving hurriedly along the walkway. Then I heard the sound of the doors of the empty cells being hastily closed and locked. My flesh began to crawl. I could hear the intruder breathing heavily outside my door. Looking down, I saw in the faint glow from the moon the doorknob beside me turning very slowly; then someone tried to open the door. The door was locked and I stared in fascination as the doorknob turned back into place and my midnight caller moved stealthily on to lock and check the remaining cells beyond me. I let my breath out slowly and deliberately. Was it the night watchman? But why? In the distance I heard talking and laughter, and as the sound increased, I realized that a group of Japanese was entering the prison compound. When they burst into the courtyard, I knew from their shuffling and loud talking, punctuated by raucous laughter, that they were drunk. One of them roared, *Djaga!* "Watchman!" The night watchman ran past, answering, *Ja, Tuan,* "Yes, sir." I broke out in a cold sweat when I heard the Japanese ask where the women from Kampili were.

"Sir, they were all taken back to Kampili," the night watchman quietly assured him. Then ensued an argument among the Japanese. I heard the word "America," which stabbed me like a dagger. Did one of them know I was still here?

---

*When I shared this with the late Dr. A. W. Tozer—the modern-day mystic, as he was called—he said, "Girl, did you ever think that God could have sent an angel?"

"Yes, Dr. Tozer, indeed I did."

Once more the night watchman averred that the women had been returned to Kampili. The Japanese were not so easily convinced. At their insistence the night watchman began opening cells, beginning at the far end of the block. The drunks spread out, yanking at cell doors in an attempt to tear them from their hinges.

My door began to be shaken. I could hear heavy breathing and smell the acrid, offensive odor of perspiration and drink. I felt faint, lest the mere beat of my heart betray my presence. Noiselessly I slid to the floor to conserve my strength, for I had every intention of doing battle. My hands were clasped tightly over my mouth to keep me from crying out.

I heard the night watchman reiterate, "The women have been returned to Kampili. Look, the cells are empty." Someone believed him or tired of the exertion, for one of the group snapped out an order in Japanese and, with a burst of bold laughter and ribald song, they left.

As their progress into the night died away, I considered the possibilities. They could inquire at the prison office of Kempeitai headquarters and learn that I was still on death row, then return; they could find more liquor and drink themselves into a stupor; or upon becoming sober later and learning of my presence, they could return to beat or kill the Indonesian watchman for having thwarted them.

Though grateful to God for my deliverance, the terror of the night and the gravity of the watchman's situation were too critical to allow sleep. The burden of my intercession was on behalf of the watchman, that God would protect him and obliterate from the memory of those lecherous, malevolent reprobates all recollection of the events of that night. Thank God, Saartje and the others from death row had returned to Kampili, for their sakes as well as mine. Though he must have known I was aware of the ruckus and the part he played in my deliverance, the watchman never acknowledged my thank-you whispered from the transom as he passed beneath. Perhaps he trusted no one either. His was a courageous act that put him in a position of extreme peril.

The twenty-seventh Psalm was a great comfort to me that night. I had committed it to memory at Benteng Tinggi. Every verse was apropos, but especially verses three and eight. "Though an host should encamp against me, my heart shall not fear: though war should rise against me, in this will I be confident. . . . When thou saidst, Seek ye my face; my heart said unto thee, Thy face, Lord, will I seek."

In the soft light of the early morning hours, I repeated these verses aloud. I had spent a restless night wondering if the drunks would return. I knew that without God, without that consciousness of His presence in every troubled hour, I could never have made it. "Lord, don't ever leave me or forsake me. Your wonderful presence has made this cell a place of beauty, a sacred place like a chapel lighted by Your presence."

Quite suddenly and unexpectedly, I felt enveloped in a spiritual vacuum. "Lord, where have You gone? What have I said or done to grieve You? Why have You withdrawn Your presence from me? O Father—" In panic I jumped to my feet, my heart frantically searching for a hidden sin, for a careless thought, for any reason why my Lord should have withdrawn His presence from me. My prayers, my expressions of worship, seemed to go no higher than the ceiling; there seemed to be no sounding board. I prayed for forgiveness, for the Holy Spirit to search my heart. To none of my petitions was there any apparent response.

I sank to the floor and quietly and purposefully began to search the Scriptures hidden in my heart. "Thy word have I hid in mine heart, *that I might not sin against Thee*" (Psalm 119:11). God's Word had always been "a lamp to my feet and a light to my path" (Psalm 119:105). I was aware that if I regarded iniquity in my heart, He would not hear me. I knew of no unconfessed sin in my heart. I believed I John 1:9, that "if we confess our sins, He is faithful and just to forgive us our sins, and to cleanse us from all unrighteousness." I knew that my sins had been blotted out, not to be remembered against me (Isaiah 43:25). Did not I John 3:21 state that if our hearts do not condemn us, then we have confidence toward God? My heart did not condemn me, and my confidence was in the Person of my Lord, Who never lies, Who is faithful to His Word. I quoted Numbers 23:19: "God is not a man, that he should lie; neither the son of man, that he should repent: hath he said, and shall he not do it? or hath he not spoken, and shall he not make it good?"

"Lord, I believe all that the Bible says. I do walk by faith and not by sight. I do not need to *feel* You near, because Your Word says You will never leave me nor forsake me. Lord, I confirm my faith; I believe." The words of Hebrews 11:1 welled up, unbeckoned, to fill my mind: "Now faith is the substance of things hoped for, *the evidence of things not seen.*" The evidence of things not seen. Evidence not seen—that was what I put my trust in—not in feelings or moments of ecstasy, but in the unchanging Person of Jesus Christ. Suddenly I realized that I was singing:

> When darkness veils His lovely face,
> I rest on His unchanging grace;
> In every high and stormy gale,
> My anchor holds within the veil.

> On Christ, the solid Rock, I stand;
> All other ground is sinking sand,
> All other ground is sinking sand.

I was assured that my faith rested not on feelings, not on moments of ecstasy, but on the Person of my matchless, changeless Savior, in Whom is no shadow caused by turning. In a measure I felt I understood what Job meant when he declared, "Though He slay me, yet will I trust in Him" (13:15). Job knew that he could trust God, because Job knew the character of the One in Whom he had put his trust. It was faith stripped of feelings, faith without trappings. More than ever before, I knew that I could ever and always put my trust, my faith, in my glorious Lord. I encouraged myself in the Lord and His Word.*

The Lord began speaking to my heart from 2 Corinthians 1:10: "Who delivered . . . and doth deliver . . . He will yet deliver."

---

*This was as valuable a spiritual lesson as the war taught me, and made even more meaningful by Dr. A. W. Tozer, with whom I shared the experience. I had been reluctant to relate this incident after having Psalm 66:18 thrown at me several times with this added jewel of wisdom: "No, there must have been some hidden sin or you would have been conscious of His presence."

"Girl, we have some spiritually ignorant people today," Dr. Tozer commented. "We are engaged in a spiritual conflict. Do these people not realize it took the Lord twenty-one days to get the answer through to Daniel because of the opposition of the Prince of Persia, Satan? God sent Michael to fight against and overcome the enemy. It is vital that we understand that Satan has accelerated his activities as prince of the power of the air because he knows his time is short."

"A wonderful verse, Lord," I responded and quoted the verse in its entirety: "Who delivered us from so great a death, and doth deliver: in whom we trust he will yet deliver us."

Immediately a still, small voice within me spoke again, "Who delivered . . . and doth deliver . . . He *will yet deliver.*"

I knew that it was the Lord speaking to me, and with joy I answered, "Yes, thank You, Lord, I'm free—free from the law of sin and death!"

Like a wave ever returning upon the beach, it washed over me: "Who *delivered* . . . and *doth deliver* . . . He *will yet deliver.*"

"Lord, I know I believe that. I'm free. I'm alive in Christ; I have been delivered."

Insistently His words echoed and reechoed, "Who delivered . . . and doth deliver . . . He will yet deliver." Finally with great trepidation I asked, "Lord, how could You get me out of here?" My mind began to explore the possibility that the Lord might be trying to make me understand that He was yet to deliver me from the Kempeitai prison. "He will yet deliver!"

I arose early, took the pins out of my hair, shook it, gave it the usual combing with my fingers, then pinned it in a roll around my head.

With great care and thanksgiving, I peeled the black, shriveled skin from *the last banana*. Eating it was somewhat of a ritual—I took small bites and chewed them slowly, savoring the flavor. I sat contemplating the luxury of God's supply when I heard the guard at the cell door. "Get up, quickly. We're taking you somewhere else." I followed him down the walkway, across the courtyard, past the interrogation room, and through the portico. The guard moved off to collect Miss Kemp and Miss Seely, so I slipped into the office. The same young Indonesian woman sat at the desk. "I'm to have my rings and Bible." She took a paper bag out of a cupboard behind her and handed it to me. I hid the rings in my brassiere and my Bible in the folds of my voluminous skirt. Stepping outside, I saw the black Kempeitai limousine waiting on the circular drive. I was motioned into the back seat. A guard got in on either side of me. By this time Margaret and Philoma had been brought and were ordered into the front seat. Poor Margaret, her condition was appalling; but Philoma—with that very un-Philoma grin on her face—made me feel physically sick.

To return to Kampili, we should have turned left at the main road. Instead we turned right, and the car continued on to the secret police headquarters. An officer motioned us to follow him. Margaret and Philoma were soon brought into the room where I was waiting. We were each given a plate of the whitest . . . the longest-kerneled . . . the most beautiful rice I had ever seen, and a small piece of boiled pumpkin. I had heard about "last meals" and paused to wonder why they even bothered.

The three of us were then summoned to a room across the corridor. On a table I saw a sheet of paper, a pen, ink, and stamp pad. While the Brain looked on, the Interrogator ordered one of us to sit down and write what he dictated. I glanced at Margaret and she nodded: "Darlene, you write it. Write whatever he tells you."

He dictated a statement of "Many, many thanks to the Imperial Japanese Army for their goodness in forgiving us our evil deeds against them." (Exactly what these "evil deeds" were was not delineated.) We promised, at the Interrogator's dictation, never again to have contact with anyone outside the concentration camp. Each of us had to sign this ambiguous statement and put a thumbprint under our signature.

How cunning! No sooner was this document made official than I found myself against a wall at the place of execution facing the Brain. He held a sheaf of papers—his handwritten record of the hearings and their accusations against me. He began to read them in Japanese. The Interrogator interpreted them to me in Indonesian. With his compilation of lies he harangued me, "You've done this and this and this, and as an American spy, you are worthy of death." With that, he drew his finger across his throat, then slapped the hilt of the sword at his side.

All time froze around me. All motion and life were suspended in that moment. In terror I watched the man's hand fold around the hilt of the sword.

Suddenly my heart burst into song: "I'll live for Him Who died for me!" I felt confused by the message of the song. "Lord, is this the deliverance—deliverance into Your presence by way of the sword? Isn't this a strange song when I'm going to die?" My heart continued to sing: "How happy then my life will be."

With mesmeric fascination I watched the Brain slowly withdrawing the sword. At that instant a car pulled up in front of the headquarters, its brakes screeching. There was yelling and running, leather heels clicking on the ceramic-tiled floors. Evidently someone yelled to the Brain, for he disappeared. I could hear excited talking, arguing. He reappeared, grabbed me by the arm, and propelled me to the front of the building. I was ordered into the waiting car next to the driver; Philoma was put in next to me. Two soldiers jumped in behind me on the drop seats. Margaret was seated behind the driver, with the Brain and the Interrogator in the back seat.

We started down the road toward Kampili as if I were top-secret material. I looked down to find two bottles of wine in my lap and somehow knew I was supposed to give them to Mr. Yamaji. It all occurred so quickly that I had no idea what had happened. I sat looking at the road, hardly seeing it. I felt emotionally drained.

"Lord, what happened?" From deep within me a verse surfaced, a verse I didn't remember having memorized. "The wicked flee when no man pursueth."

The joyful anticipation of soon escaping from the Brain and the Interrogator was short-lived. No sooner had we crossed the moat to enter Kampili than I felt my emaciated left arm clutched in an iron grip and twisted. The pain was so excruciating that I bit my lower lip to keep from screaming. "Dear God, he's going to break it!"

The Brain hissed, "If you ever have contact with anyone outside the camp, I'll get you; or if you ever tell anyone about what happened to you, you'll not escape the next time!"

Terrible fear came over me as the car slowed to a stop in front of the commander's office. The officers and soldiers jumped out and walked inside. Sweet Seventeen, who had come out to greet us, accepted the wine for Mr. Yamaji.

Margaret, Philoma, and I looked and felt very confused. We hadn't been dismissed, but as a few women cautiously approached, I asked if they would take Margaret and Philoma to the hospital. Ruth was coming from the sewing room, and I started toward her. With an arm around my waist, she guided me to Barracks 8. At her mirror hanging at the foot of the bed, I stopped to see how well I'd done at combing my hair with my fingers. In shock I cried, "Ruth, look at my hair; it's white!" I saw her tears as she nodded and turned away. Lilian had seen Margaret and Philoma, then came to express with a hug, "Thank God, you're back!" She and Ruth didn't look in much better condition than I.

Before I could greet the others who were waiting, a woman called, "*Mevrouw* Deibler, I must speak to you immediately." I followed her into the dining shed. Pulling from her pocket a washcloth with "Darlene" embroidered across it, she said in a stage whisper, "I was in prison in Macassar—another prison— with a friend of yours Wiesje Kandou. She asked me to give this to you. Here, take it!"

I drew back as if she were handing me a death adder. The Brain's recent threat echoed in my ears. I remembered that it was this woman who had cried so vociferously to be taken back to Macassar the day the Japanese had stopped at Benteng Tinggi en route to Malino.

"No," I cried, "I'll not take it!"

Her face contorted with fury as she shrieked, "You take this! I risked my life to bring it to you!"

I backed away from her, crying, "No! I'll not take it! You're trying to incriminate me. If you don't take that and leave here immediately, I'm going to Mr. Yamaji and tell him. I know he'll believe me."

Seeing Ruth enter the shed, I sobbed, "Thank God you're here!" Then I told her what had happened and of the Brain's threat.

Ruth turned on the woman: "You get out of here, and don't you ever come near Mrs. Deibler again!"

The woman fled as she became aware of women closing in around her. "She was in prison? Bah! She was gone only a few days!" one woman exclaimed.

I had to lie down. What strength I had was gone.

Fear gripped me like a physical thing that I couldn't shake off. Questions paraded through my mind. What will the Kempeitai try next, since this ploy failed? Where did the woman get the wash-cloth, thread, and needle—if not from the Kempeitai? Was the Kempeitai waiting for a signal from her, that the washcloth was in my hands? Would that explain why she was in such a hurry to get me to take it? The "death wagon" could now be heard leaving.

Fear that I would fall into the hands of the Kempeitai stalked me by day and held me in its clutches by night. I was afraid to go to sleep for fear my mind would go, so awful was the terror by night. I didn't dare to unburden my heart to anyone because of the Brain's threat of what would result if I told what had happened in prison.

By day I worked wherever I was needed; by night I prayed for deliverance. The afternoon of the sixth day after my return, I walked out onto a grassy plot in front of our dining shed. "Lord, I've sought You for deliverance from this terrible fear; I need sleep, but I'm afraid to go to sleep. I have no more strength left!" In despair I threw out my hands, crying, "Lord, I'm gone. I'm gone!" In that instant I felt arms go underneath me and found I was singing the third verse of a song by Dr. A. B. Simpson:

> Underneath us, O how easy!
> We have not to mount on high,
> But to sink into His fullness,
> And in trustful weakness lie;
> And we find our humbling failures
> Save us from the strength that harms;
> We may fail, but underneath us
> Are the everlasting arms.

As I sank into His fullness, the Lord completely delivered me and all fear was gone. "For God hath not given us the spirit of fear; but of power, and of love, and of a sound mind" (2 Timothy 1:7).

CHAPTER NINE

Margaret Kemp's memory was not improved, but with time the bruises disappeared from her arms, back, and legs and after several weeks the doctors felt that she was physically well enough to return to her work in the sewing room. Philoma's physical problems had been taken care of, but her mental condition was such that she needed either to be confined where she couldn't wander off, or be put in restraints. In the heat of the tropics, restraints would have been unbearable, so the doctors confined her in a small hut on the edge of the camp. The room was large enough for her to move about in. The door was kept locked, but there was a barred window where she could look out. When I went down to see the hut, I noticed the chicken pens nearby. It came to me that if I were allowed to work with the chickens, I would be near Philoma and could visit with her before and after work. (I had not resumed my work as barracks leader. Rose David was doing very well, and I had been filling in for those who were sick, whatever their work assignment had been.)

I went to see the nurse and the artist, who were responsible for the chicken pens. They were very gracious, and neither Mr. Yamaji nor Mrs. Joustra objected, so I began working regularly with the chickens.

I enjoyed the work immensely. One mother hen, with about a dozen baby chicks, became a real pet. Each morning we let the chickens outside the pens to forage. When it was time for us to leave in the late afternoon, we trilled for them to return to the run. The mother hen stood clucking until all the chicks gathered about her and then led them off into their pen.

One morning soon after I began working with the chickens, the nurse and I arrived early to find that the mother hen was not out in the run. She was sitting inside the shelter with her chicks still under her wings. Seeing us, she stood up and said in "hen language," "All right, let's move out." When they got to the door of the run, we noticed that the hen's neck was bare and raw. It was evident that something had torn her neck; the skin held together in only one small place. A quick count showed that all her chicks were there. Somehow she had fought off the attacker, but she had nearly lost her life protecting her little brood. Immediate surgery was indicated, so the nurse went to her barracks for some balsam of Peru and a needle and thread. By turns we pulled the skin together and stitched it, all the while crooning to her what a good, brave mother she was. Each time we stuck in the needle she cried, "Awk!" but never once struggled to get away. We covered the whole neck with balsam of Peru to keep off the flies and preclude infection. The surgery was successful and the patient survived.

While the chickens foraged, we collected the chicken manure on bamboo mats to sun-dry. We pounded the dried manure in a large wooden mortar with a six-foot-long pestle. With our hands we scooped it out of the mortar and into the sieve, making sure that no chunks remained. The powdered fertilizer, once bagged, was sent to Macassar for the Japanese flower gardens. Knowing that we could collect the eggs for the hospital, to be given to the critically ill, made our task far from onerous.

One noon, returning from the barracks, I stopped, as I usually did, to talk to Philoma. She was holding onto the bars of the window, swaying back and forth with her eyes closed. I looked at her gray hair, uncombed, hanging down her back. She was partially clothed, but when I became aware of what she was saying, it broke my heart. What indignity and pain she must have suffered. I turned quickly away and dropped onto a log near the hut. Putting my hands over my ears, I began to sob, "O Lord, how can this be? How could this happen to Philoma, a woman of such faith? Why did she have to suffer such terrible humiliation?"

Suddenly I felt a touch on my shoulder. It was Mr. Yamaji on his bicycle. I don't know when he came or how long he had been there. I jumped up, desperately trying to wipe away my tears.

"Don't cry, *Njonja*. She doesn't know what she's doing or saying. What's more, why are you grieving like this? You trust in the Lord." That rebuke, coming from him, was exactly what I needed.

"I'm sorry, *Tuan*. You're right, but it's terrible to see *Nona* Seely like that." He sat for a moment, looking off in the distance, nodding his head, then rode away. I went back to Philoma, placed my hands over hers, and softly called her name. She became quiet as I prayed for her. She looked at me and smiled. I knew she recognized me. I returned to my work, wondering if Mr. Yamaji really understood what it meant to trust in the Lord.

When Philoma became mentally stable enough to be allowed to move into one of the stone houses with Margaret, Mrs. Jaffray, and three other women, my reason for working with the chickens—to be near her—no longer existed. Philoma helped care for more than two hundred ducks outside the boundary of the camp.

My heart was at rest concerning Philoma's progress, so when Rose David asked me to resume the leadership of the barracks, I agreed. Rose wasn't feeling well and the doctor suspected a heart condition. I had gained a little weight and felt better. I'd continue to fill in for the sick and function as barracks nurse.

In the early months of 1945, there was increased air activity on all sides of the camp. More and more nights were spent in the slit-trenches we had dug. Daytime raids became more frequent. We witnessed a second dogfight just beyond the camp. A Japanese plane was shot down and crashed in flames. We had learned, the hard way, not to vocalize our cheers.

Night after night, from our trenchside seats, we watched the Allied planes, caught in the searchlights, coming in to drop their bombs. Desperately we prayed that the planes would escape the antiaircraft shells bursting around them. When we saw them slip through the "ack-ack" without breaking formation and apparently unharmed, we slumped down with thankful hearts and closed our eyes, waiting for sleep or the next bombing raid to come. Even if it rained during a raid, we dared not return to the shelter of the barracks. Seventeen inches allotted to each person in the trenches didn't permit reclining or moving about. By morning we were desperately weary.

Those were suspense-filled, nerve-racking nights. We became quite adept at telling time by the position of the Milky Way and the Southern Cross. I often thought of Genesis 1:16: "He made the stars also." No room for doubt as to the origin of those stars; they are the work of His hands. I thrilled to their beauty, but I worshiped their Creator. We waited with anticipation for the rising of the magnificent, bright morning star, even as we awaited the coming of the bright Morning Star, our wonderful Lord Jesus.

Daytime raids also became more frequent. One afternoon in mid-July, when all the school children and most of the women were out of the barracks working, a plane flew low over the edge of the camp—so low, in fact, that we could see the pilot in the cockpit and the American flag and insignia on the fuselage. It was as if the crew had avoided coming directly over the camp, so we felt that they knew it was a camp for women and children. However, when the plane circled and came in again, even lower, we stood awestruck and mute watching it. Then, in an open area of the camp, the pilot dropped a large metal object, turned, and took off into the wild blue yonder, waggling the wings. The object didn't explode, but what was it? I felt anger rise in me, thinking how easily he could have hit one of the children. Was he blind or drunk? Couldn't he see the women and children? The object proved to be an auxiliary fuel tank.

What was going on? What was the message conveyed by dropping an empty fuel tank? Never in the more than three years of our stay had we received letters from home or Red Cross packages, nor had any pamphlets found their way into the camp to tell of the progress of the war. Each year there had been a series of air raids. Sometimes it seemed we would spend the rest of our days locked behind the barbed wire. Work on the air-raid shelter, between the hospital and the camp commander's office, had been accelerated. Yet we avoided being inquisitive about anything where native workmen were involved. So the auxiliary fuel tank was wrestled out of the way, and normal camp life was resumed.

Two days later, early in the afternoon of July 17, we heard the sound of many planes. The air-raid alarm was sounded. We all ran for the trenches and sat to watch the planes. They were headed straight for our camp. The planes were silver and had double fuselages, and they were dropping silvery objects. Some of us were yelling, "Chocolate bars!" Others were crying, "Canned goods!" Still others screamed, "Pamphlets!" All of us were seeing illusions of the things we longed for most—food and news! Our joy was shortlived, for suddenly we recognized the whistling of bombs; in minutes our whole camp was in flames. I jumped back into the slit-trench, but the second my feet hit the bottom of the trench, the Lord said to me, "You borrowed Mrs. Lie's Bible!" (Mine had fallen apart.)

"You're right, Lord. I have no right to let her Bible be burned!" Jumping out of the ditch, I ran into the burning barracks, scrambled up the ladder, grabbed her Bible off my bed, dropped to the floor, and ran outside to get out from between the burning barracks and the dining shed, not knowing in which direction they would collapse. I saw Rose David standing in front of me.

"Darlene, where do we go?" she cried. Wherever we looked, there was fire.

"I don't know, Rose."

"Oh my God, if you don't know, who does?"

Just then I saw that the gate leading out of the camp had been opened; women and children were fleeing across the moat and over the embankment. "Rose, the gate's open. Come, run!" As we neared the gate, I saw that it was the Brain who had opened it and was standing guard. My insides turned to water: "No, Lord, please don't let him recognize me!" I turned my face away and ran, trusting that he wouldn't notice me.

No sooner had we crossed the moat and the road than we realized that the camp was surrounded by hundreds of Japanese soldiers with their machine guns set up on the embankment ready to fire on the planes as they circled to make their second bombing run on the camp. (We learned later that there were thousands of Japanese soldiers who had withdrawn from the other islands to make their last stand on the island of Celebes.)

Running down into the midst of the soldiers, we thought only of passing through as quickly as possible into the surrounding rice paddies, where we saw others of the women and children.

Tidur! "Lie down!" screamed the Japanese as they turned on us with their guns, and *tidur* we did. Not another step. Wherever we happened to be, we threw ourselves prostrate on the ground. A bayonet jabbing at my back was a great persuader. If we were in the soldiers' way, they ran over us and began firing at the planes. The planes immediately circled, coming in low to strafe the Japanese and knock out their machine guns. Bullets were flying all around as twenty-three planes passed over. How could the stray bullets miss us? I braced myself, knowing that it was altogether possible I could be hit. Dropping my head onto my arm, I whispered, "Lord, if anyone is alive at the end of this day, it will be a miracle."

God wonderfully ministered to me, through a song, in this time of terror.

> Other refuge have I none;
> Hangs my helpless soul on Thee.
> Leave, O leave me not alone,
> Still support and comfort me.
> All my trust on Thee is stayed,
> All my help from Thee I bring;
> Cover my defenseless head
> With the shadow of Thy wing.

Suddenly the shelling stopped and no one was moving. Looking up, I could see the planes disappearing in the east. "God, it's a miracle, I'm alive! I'm alive! Thank You for overshadowing me!"

The Japanese burst into activity. Grabbing their machine guns, they disappeared across the rice paddy. As I stood up, I saw Ruth with other women and children from our barracks emerging from their hiding place beyond the long grass. Ruth was carrying her Bible and a small bag of things she had prepared and had been able to retrieve from the barracks before it had been hit. I suggested that we return to see if we could find our spoons, tins, and other belongings in the ashes. We could see nothing of Margaret Kemp or Lilian. We crowded through the gate and returned to where Barracks 8 had been.

I stopped in front of where my bed would have been when it dropped to the ground burning. There, on the top of the heap of ashes, lay my Bride's Book—my beautiful Bride's Book that I had carried with me all these years, sewn inside the native sleeping mat. Somehow—*no, not "somehow," but by my Father's ordaining*—the fingers of the flame had peeled away the mat and flicked through the pages to the centerfold, where my marriage certificate was written in gold ink. I gasped . . . it was so beautiful, that bright shining, gold ink on the black, charred page—gold, purified by fire, glittering in the rays of the late-afternoon sun. I dropped to my knees and reached out, but the moment I touched the book, it disintegrated and was gone. I rocked back on my heels and in anguish cried, "Lord, that was the only thing I had left! Couldn't I have had that? Just that one thing?" I covered my mouth to keep from screaming. I closed my eyes and crooned, "Father! O Father!"

Gently, so gently, He answered me, "My child, that's what I want to do with you—make you like pure gold—even if I have to take you through the fire seven times."

I was shaken to the depths of my being when I absorbed the enormity of what He had said to me. "O Father, seven times? I don't have anything left to give You . . . but myself." I felt His arms of love lifting me up. I stood to my feet, slashing at my tears, for I heard Ruth calling to me. She had found a green papaya that had cooked in the fire. She broke it open and, with our hands, we stripped off the charred skin and ate. Sifting through the ashes, I found a small teaspoon and a dessert spoon, birthday gifts from friends of my barracks; the watch Russell had given me; and three lockets, gifts from the Jaffrays. By this time, we were very dirty from the juice of the papaya and the ashes. There was no water to be had from the well, as the ropes on the buckets used for drawing water had burned. As best we could, we wiped our hands and faces on our work suits.

I heard someone sobbing. Looking around, I saw the leader of Barracks 7 crying. I went over and put my arm around her. "My mattress burned," she sobbed. It was just a thin pad she had been allowed to keep because she was an older woman.

"Oh, yes, everything has burned, but we're still alive. We have much to thank God for!" I reminded her.

"But I didn't leave it in the barracks. I threw it in the ditch where you always lie."

I felt the hair stand up on the back of my neck, and a chill went through me. Walking over to the edge of the ditch, I looked down. There, where I had been crouching, was the casing of the bomb and the ashes of her mattress! I walked away, unable to talk, so great was the sense of awe that had come over me. "Lord, it wasn't Mrs. Lie's Bible You were concerned about, was it? You knew that was a way to get me out of that ditch . . . to save my life! Father, whatever is left of life to me, it's Yours. It all belongs to You!"

"*Mevrouw* Deibler! *Mevrouw* Deibler! Freddie's been hit. They're taking him to Macassar!" one of our boys called. Ruth and I ran over to the group of young people. They all began to talk at once, telling how they had been in the trenches at the far side of the camp with one of their teachers, an older woman, when the planes began to drop the bombs. Freddie had been lying face down next to the teacher, who, on hearing the planes approach, had thrown herself across the upper part of his body. However, the bomb hit Freddie, passing through his right buttock and almost completely severing the right leg from the torso. The bomb had not detonated, thank God, or all of them would have been sprayed with the gasoline jelly, and, in all probability, burned to death. The vehicle that had just left the camp was taking Freddie to Macassar for surgery, along with another young woman whose leg had been severed below the knee. Freddie's mother, Serah, had not been allowed to accompany him. I felt sick with grief for Serah and Dolly, Freddie's sister. We stood and looked at one another, weeping silently.

The doctors and nurses, who had sheltered with the hospital patients, Mr. Yamaji, and the second-in-command in Mr. Yamaji's bomb shelter, were going through the ditches, giving sedatives to the burn patients. Then they immersed them in the long cement hospital water tanks, in an effort to ease the pain. There was no screaming or loud crying. After more than three years, we had become a people who were no strangers to suffering. "Dear Lord, comfort these people; ease their hurt," I prayed silently.

The stillness was broken by Japanese soldiers gesturing and yelling, *Pigi! Pigi!* "Go!" Go? Go where? All the bamboo structures had burned, and much damage had been done the cement houses. Waving their arms, they indicated that we should start moving in the direction of the area where Barracks 1 through 6 had formerly stood. The barbed-wire fence had been cut. We were herded through the opening, across a rice paddy, and up into the surrounding jungle. Much to our amazement, there, hidden among the tall trees, were new bamboo mat-walled, grass-roofed huts!

"So the Japanese had been *expecting* our camp to be bombed!" we said to one another. Little clearing had been done in the area, so the huts would not have been conspicuous to us from the old camp, or to the Allies from the air.

We found the one-room shack designated for Barracks 8. Fortunately the huts had been built up off the ground, because it was very damp in the jungle. Emotionally drained and physically exhausted, we crawled inside to sit and wait for families and friends to find one another. Margaret Kemp and Lilian were there waiting for us, looking very pale and weary. They each had a little bag of items that they had carried to work with them each morning. I looked about me, trying to decide how we were all going to fit in this limited space. At least we would have a little more room than in the trenches. Maybe we could stretch out to sleep. That would be a blessing. The floor was made of two-inch-wide bamboo strips tied together with rattan, about an inch apart, providing easy access for the wind, the mosquitoes, and all the other "creepy-crawlies" of the jungle. If the floor had been designed for a bed of torture, the builders outdid themselves. There was spring to the bamboo, but it was no "beauty-rest."

Someone passed the word that the kitchen crew was preparing something to eat. Rice, vegetables, and whatever could be salvaged from the burned-out kitchen had been dumped into drums and cooked with water to reappear as a vegetable gruel. We were grateful for whatever was available, and we had long since learned that our cooks, under Mrs. de Jong, had the genius of making whatever they cooked tasty.

Having eaten, we set aside our tins, plates, or whatever had been retrieved from the ashes and wandered back to use the now-open-air latrines. Water was hard to come by, so we shrugged our shoulders and said, "No water, no towels . . . so we bathe tomorrow—or whenever!"

Margaret Jaffray, her mother, and Philoma had lost none of their possessions. Their stone house had sustained some damage but hadn't burned. Margaret cut her blanket into three pieces to share one-third with Margaret Kemp and Lilian, one-third with Ruth and me, saving only one-third for herself. We were very grateful to her, as all our blankets had burned.

Knowing it would soon be too dark to see, I suggested that we go inside and mark out our territories. It was important to have those who had to get up at night and families with small children near the door, in case of necessity or an air raid. There were now no lamps and no bamboo pots.

Never had our hearts been so united in prayer as that night. I had Mrs. Lie's Bible, but it was too dark to see, so I began quoting one of the Psalms that had been most often on my lips during those difficult years, Psalm 27.

> The Lord is my light and my salvation: whom shall I fear? The Lord is the strength of my life; of whom shall I be afraid?
>
> When the wicked, even mine enemies and my foes, came upon me to eat up my flesh, they stumbled and fell.
>
> Though an host should encamp against me, my heart shall not fear: though war should rise against me, in this will I be confident.
>
> One thing have I desired of the Lord, that will I seek after; that I may dwell in the house of the Lord all the days of my life, to behold the beauty of the Lord, and to enquire in his temple.
>
> For in the time of trouble he shall hide me in his pavilion: in the secret of his tabernacle shall he hide me [vv. 1–5].

"Lord, if ever we needed a place to hide, we need one now!" There were subdued sobs here and there as I finished the Psalm and prayed for the bereaved—and especially for Freddie and his family—that God would ease their pain and suffering. "And please, Lord, for the sake of the children, let there be no air raid tonight. There's no moon and we don't know this area, so where would we hide? Please, Lord, don't let the planes come tonight!" The amens, like a gentle wave, swept through the building.

It being "the year of the very lean body," our thinly padded bones found little comfort, no matter how we twisted or turned. Most of us were wearing our work shorts and sleeveless blouses. Without nets, we had no protection from the myriad of mosquitoes. My legs and arms were on fire, and from the scratching I heard, I knew others were suffering the same discomfort. The mewling of the little ones was distressing, but thanks be to God, there was no air raid, not even the distant sound of planes, to disturb them. Finally the heavy breathing told me that most of the people had drifted off into that deep sleep that follows total physical and emotional exhaustion.

There, in the dark hours of that night, I walked into the sanctuary of my heart. The lamps fed by the oil of the Spirit were burning brightly. "My precious Lord, I have come to worship and adore You. This has been a day like no other I have ever known. Today has marked the final stripping away of every transient treasure I possessed. I've nothing but this dirty, faded blue-gray work suit, but never have I felt so privileged, so blessed, or so rich!"

I thought of the many nights spent in the trench, looking up at the night sky. I reveled in the magnificent display in the heavens— the stars, the moon, and the planets—and I wondered how such a One, the great Creator, could have a personal interest in me, a young woman without any special gifts, talents, or beauty. Sometimes the very magnitude of His handiwork made Him seem almost remote. But that night, that One, the High and Lifted Up Holy One of God, wearing His most magnificent robe, a robe of human flesh, came to dwell with a child of man in a new and beautiful relationship. Oh, the wonder of His love for me and His personal concern for me, as an individual, was overwhelming.

Together we walked through the events of that day. I heard again the insistence in His voice, as He reminded me of Mrs. Lie's Bible—not my Bride's Book, nor my full five-year diary—it had to be Mrs. Lie's Bible that was lying on my bed, directly above the ladder. Then it became clear to me why He didn't remind me of either of the two books, neither of which could ever be replaced. I would have been up there, tearing at the mat trying to retrieve them and, without doubt, the burning building would have collapsed on me. I just escaped, as it was. "O Lord, You saved my life! Thank You for reminding me of this Bible!" I hugged it to me, for Mrs. Lie had said I should keep it.

I had much to be thankful for. Of all the Japanese I had ever had any contact with, none terrified me like the Brain. I poured out my gratitude to God that the Brain hadn't noticed me when I passed through the gate. I thanked Him for the miracle that none of us was killed or hurt during the machine-gun strafing.

I knelt again before my Bride's Book, lying open on the heap of ashes. I saw how bright and shining the gold ink had become. The gold had to pass through the fire to destroy the tarnish, and it was the background of a black, charred page that displayed its beauty—a beauty I had never noticed, written, as it had been, on a bright white page, surrounded by pretty flowers! "I understand, Lord, I really do understand what You're saying to me through this. Forgive my tears!" I learned that my tears were a gift from Him to ease the hurt, a gift to be shared with others who were hurting. Was I not to weep with those who weep? (Rom. 12:15). "When the tarnish begins to appear, take me through the fire, Lord. I'm available."

One by one, He pulled Scripture passages out of the storehouse of my memory, to remind me that they had been hidden there for just such a time as this. "I have chosen thee in the furnace of affliction" (Isaiah 48:10). This verse called to remembrance my Lord in a blazing furnace with three young Hebrew men. There was something so poignant, so intimate about the privilege that was theirs, of "walking around in the fire" with their Lord. Because of the testimony, the faith, and the courage of three young men, the king and a crowd of people caught a glimpse of Jesus. When they emerged from the furnace, there was no acrid, caustic scent of fire upon them—just the fragrance that emanated from three young people who had been walking with their Lord in the furnace of affliction.

"That's very important, isn't it, Lord? I pray that, if I come out of this war alive, I may be "sweet-smelling"—not bitter or cynical, but like a sweet-smelling, fragrant incense unto You. All this long day, You have walked with me, and never for a moment have I been out of Your sight. Of this You have made me keenly aware."

I saw again, in my mind's eye, the bomb canister there in the ditch among the ashes of a mattress, and I knew how much my Lord loved me. Singing what has come to be my Lord's lullaby for me, I fell asleep:

Loved with everlasting love,
Led by grace that love to know;
Spirit breathing from above,
Thou hast taught me it is so!
Oh, this full and perfect peace!
Oh, this transport all divine!
In a love which cannot cease,
I am His and He is mine!

At first light the next morning, Ruth found herself in full possession of our mutual piece of blanket. "Oh, I'm sorry. I guess I didn't realize how narrow it was," she apologized. "That's all right, Ruth. You flounced sooner and stronger than I did, and I didn't have the heart to wake you up." It was good to be able to laugh. From then on we fell asleep back to back, with the piece of blanket covering our topsides. Whoever flounced first usually stayed in possession of it till morning.

Blankets weren't the only thing in short supply. With an average of one comb to nine people, and head lice on the increase, Ruth and I decided that the wisest course was to cut our hair off short. Ruth had her scissors, so she cut mine and I cut hers.

Most of the women were detailed to work in the garden and the old camp kitchen. The chickens and pigs had been incinerated by the fire. We dared not eat the pigs, because of the gasoline jelly; otherwise, that would have been quite a barbecue!

The planes returned three days later to bomb the old camp with shrapnel bombs, in an attempt to destroy all the stone houses and the Japanese headquarters. We hid during the bombing among the trees near our hut.

As it was growing dark the following evening, Sweet Seventeen came to say that Commander Yamaji wanted me to come. I ran across the rice field to the old camp, arriving in time to see a car pulling away from the headquarters. Word had come from Macassar to say that Freddie had passed away and would be brought in the following day. They had excised the leg completely, but the gasoline jelly had eaten into the intestines, so it was impossible for them to save him. "I want you to tell *Njonja* Paul. You have received similar news, and you will know how to help her as you have been helped. The teacher is still in the hospital. They took her leg off above the knee in order to save the thigh, but they couldn't save *Sinjo* [master]. You will tell her?"

I assured him I would, though I had never done anything like this before. "And I will pray with her and the sister, for it is God alone Who can help a person in a time like this. It will be very difficult for *Njonja* Paul; Freddie was her only son."

"I understand, but you will know how to help her. He will be brought in tomorrow; then you can tell her." He bowed to me and said, "Goodnight. Thank you, *Njonja*."

I returned his bow and left, wondering what I would say when they asked me what the commander had wanted. Freddie was the apple of Serah's eye. What could I say to help ease the pain? How sad—sad for all the family. It was a fairly long walk across the rice field to our jungle camp. When I arrived at the hut, everyone was sleeping, so I felt it best not to awaken them for devotions. I found my place near the door, next to Ruth, but I couldn't sleep for thinking of Serah and Dolly's anguish. Suddenly the sound of the air-raid alarm and the distant hum of approaching aircraft startled everyone into wakefulness. Pandemonium broke out, children and mothers crying out trying to find one another in the dark. A traffic jam occurred as they all tried to exit simultaneously through the small, one and only door. I checked to make sure everyone was out of the hut, then groped my way with others from tree to tree, seeking a place to shelter away from the bamboo buildings.

The stygian darkness enhanced the awful fear that made our pulses race and our throats feel dry. We knew that the planes would soon be directly overhead. Was our new camp the target? What kind of bombs would they use? Facing death again, we committed ourselves into His care. None was ashamed to pray for mercy. With bated breath, we waited to hear the whistling of the bombs; then, miracle of miracles, the planes passed us by—or were they checking the location of the jungle camp? No, they continued on toward the west. Just as fervently as we had prayed for mercy, expressions of thankfulness were voiced unashamedly to God. The raid took place on a Japanese airfield and encampment not far from us. No one moved until the planes could no longer be heard and the all-clear was sounded; then we made our way back to the hut.

As I took roll call to make sure no one was missing, discussion began at the other end of the hut. Finally one of the women called out, *"Mevrouw* Deibler, we need to talk to you. Every night during devotions, you've prayed that God wouldn't allow the planes to come, especially for the sake of our children, and it's so difficult to find a place of shelter during the dark of the moon. And the planes have *not* come. But tonight you weren't here for devotions. The planes came. If you're called to the headquarters again, we want you to know, we don't care what time it is when you return— please waken us so that you can pray with us that the planes will not come."

I never had to waken them; they were always waiting for me. Never again did we have a night raid, or even the sound of approaching planes, to disturb our sleep.

I missed Serah while collecting porridge the following morning. As soon as I had finished sorting out work details, I went in search of her. I met her returning from the vegetable garden and fell in beside her, slowing my pace to allow others of the garden detail to get ahead of us. When I placed my hand on her arm and stopped, I saw the color drain from her face. "O God," my heart was crying, "give me the right words to say." Then as gently as I knew how, I told her about Freddie. She reached for my hands, and neither of us could restrain the tears. "Serah, I'm so very sorry for you and Dolly. I'm sorry for all of us; Freddie will be missed. I do understand your grief. Believe me, I do." I held her and we prayed until a quietness, a peace, and an acceptance of His will gave us the strength to go back to the shack. At her request, I told Dolly and the others about Freddie's death. Lovingly and quietly, the people surrounded Serah and Dolly to share their sorrow. It was very hard for the young people to accept that Freddie was gone. They had been a close-knit group, and un-ashamedly they cried for the death of their friend. We went as a group to headquarters to wait for the ambulance that was to bring Freddie back. I suggested to Serah that I be allowed to keep the wake that night, as she and Dolly would need their rest. She agreed, knowing that Dolly wouldn't go back to the hut alone.

When the ambulance arrived, Freddie was carried into the commander's office and laid on a bamboo bed, made ready by Mr. Yamaji. Serah and Dolly went inside while we, who had accompanied them, waited outside. Finally Serah came to the door to nod that it was all right for us to come in. Silently we mourned with them. The lower part of Freddie's body was covered with a sheet. He looked very peaceful, and I was grateful that his face didn't show how very much he must have suffered. With an embrace or a word of sympathy to Serah and Dolly, the various ones took their leave. I whispered to Serah that I would be back to relieve her and Dolly as soon as devotions were finished. She nodded and I left, knowing that those closest to her and Dolly wouldn't leave until I returned.

Once I was alone with Freddie for my vigil, I set a chair near the head of the bed; a small oil lamp had been left burning. It was a night for thinking and listening. I had never been alone and this close before to a person who had passed away. I turned the chair so that I could look at Freddie's face; then the words of a very wise man came to me and the voice of memory spoke: there is "a time to be born, and a time to die" (Ecclesiastes 3:2). Both are times accompanied by pain: the pain of giving birth, mingled with the exquisite joy that a child has been born; and the pain of death, mingled with the bitter dregs of sorrow that a child has died. Serah's joy and pride in her son had turned to pain and a great sense of loss.

I thought about Freddie as he had been, a tall, handsome young man of fine physique, anticipating a happy life "when the war is over." I wondered—if he had lived, could he have coped with the loss of his limb, or might the loss have left him bitter and psychologically, as well as physically, crippled? God alone knows, and in His infinite wisdom, He had taken Freddie, who was so young, while many who had lived well and long were still among us. I only know that my Lord makes no mistakes.

I wept in my own aloneness, remembering another day and another year when I had lifted the cup of sorrow to my lips and drunk my portion of the bitter dregs. But in the dark hours of that night, He came to me, the Author of Life, to remind me of His words: "I am the resurrection, and the life: he that believeth in Me, though he were dead, yet shall he live!" (John 11:25). Once more peace and comfort, like a mighty river, flooded my soul. "O death, where is thy sting? O grave, where is thy victory!" (1 Corinthians 15:55). I prayed for Freddie's father in Pare Pare, wondering if it would be months before he learned of his only son's death I thought also of other bereaved families whose husbands, fathers, and brothers had yet to learn of their losses.

When the first streaks of dawn lightened the sky, I looked up to see Mr. Yamaji standing in the doorway. "*Njonja,* you go. I will watch." I nodded and left, praying that God would bring to his remembrance our discussion of two years ago concerning death, sorrow, hope, and life everlasting.

Few of us will forget the memorial service held in the open field in the northeast corner of the old camp. It was appropriately conducted by the Catholic priest and the Protestant dominie. The commander and the second-in-command attended in full-dress uniform. The service over, we filed silently past the coffins draped in white cloth. The victims of the bombing were buried alongside our earlier dead—the line of simple mounds lengthened, the starkness of bare earth soon to be softened in green.

A strange stillness pervaded our jungle camp. It was like a breathless waiting, but for what? We didn't know. We remembered the bombing raids of the previous year, the months of miserable nights spent in the rain in the muddy slit-trenches. After months of air raids, they had ceased, so we had relaxed and dared to entertain the hope that peace talks were in progress and the end of our imprisonment was imminent.

Then a year later, we had found ourselves in the midst of a repeat performance—months of day and night bombings around our camp, nights in the trenches, shivering in the rain. But there the similarity ceased. Our camp—a camp of women and children—had become the target for bombing raids—not just once but twice targeted for total destruction! Why? Was it to destroy a work force used by the Japanese, or were we just expendable?

We buried our dead and were then set to the onerous task of stacking the canisters from the incendiary bombs and cleaning up the old campsite. The stinking mud of the gardens aggravated our tropical ulcers. Our eyes searched the sky for approaching planes. We noted the nearest trenches as we moved about the camp, to know where to run in case of another bombing raid.

The doctors and nursing staff, with those of the sick who were not ambulatory, were reestablished in the damaged hospital. It was near the commander's bomb shelter. The planes were no longer coming over the camp, but there was daily activity to the north of it. We wondered if Pare Pare was being bombed. We were greatly concerned for the men, and nightly our prayers encompassed them with petitions for their safety.

We wondered if our camp was to be relocated, or would we be left in that dreadful jungle camp? A terrible apathy settled upon us. Animated speculation about "after the war" or "when I'm free" was no longer heard. Even the children were listless. We had no illusions about being left "ladies of leisure"; however, contemplation about the future was debilitating. To make morale worse, the grass roof was leaking, so restless nights followed rainy days.

Most of us had no change of clothing. We worked in our one set of clothes, we bathed in them in the rain without soap, and—drip-dried or not—we slept in them. "Lord, we need some kind of diversion! We're becoming a tribe of zombies." The diversion came! It came in the form of a very large monitor lizard! The young people saw it one morning and were trying to corner it under our shack. They had armed themselves with clubs and were yelling to us women inside, "Can't you see him?" Oh yes, we had no trouble seeing through the slatted floor the drama that was taking place under us. The lizard had lifted himself high on his legs; his mouth was open and the tongue was darting back and forth as he hissed, showing very sharp teeth. He was looking to see if he could break through the floor. When the clubs were thrown at him from one side, he raced to the other end of the building, trying to escape. We fled to the opposite end of the hut, to escape the teeth tearing at the bamboo and rattan.

The lizard would follow us and we'd scream, "He's here! He's here!" as we jumped up and down, trying to keep our feet from being bitten. Clubs were flying from all sides. The children were yelling as they wriggled under the shed to retrieve their clubs. We women screamed and yelled and laughed while running from one end of the shed to the other in a pack, thinking what a marvelous addition he would make to our rice and vegetables. (Lizard tail is very good eating. At that time, what *wouldn't* be very good eating?)

But alas, while the children shifted, retrieving their clubs, the lizard made good his escape from under the house and headed for the jungle with amazing speed. We darted outside and grabbed up clubs to join the chase. Racing through the trees, side-stepping some of the large ferns and dodging the hanging vines, we followed on, yelling encouragement to those in front of us. The children finally gave up, admitting defeat.

We drifted back to the shack and threw ourselves on the floor winded, but still laughing, as each related his or her part in the "Battle with the Monster"! Growing quiet, we sat up, looked at one another, and smiled. How good the excitement, the yelling, the laughter, the releasing of pent-up emotions had been. For a half hour or so, we had become savages, intent on food, exhilarated by the chase, enabled for a moment of time to forget the horrors we had experienced during the past weeks, and the familiar drudgery of today and tomorrow.

Soon the commander brought in native laborers to begin construction of a new school house on the old campsite. One early afternoon, Cor Voskuil came to say that all barracks leaders were asked to assemble in the partially completed building; Mr. Yamaji had an announcement to make. Entering the building, we found Mrs. Joustra standing in front of Mr. Yamaji. When he said, *"Njonja* Joustra is no longer head of the camp; *Njonja* Deibler is now the head," a shock wave went through the building that left us speechless. I couldn't believe what I had heard, nor could the others. I stood dumbfounded!

Recovering, I gasped, "Oh, no, *Tuan,* I couldn't take *Njonja* Joustra's place. I don't know enough about running a camp like this." Before he could reply, one of the other barracks leaders broke in, complaining that I was much too young and one of them should have been chosen. I felt grateful for her protestations. I sincerely didn't want the position.

Mr. Yamaji was angry at the uproar. His cane came down on one of the desks with a terrible crash, and there was instant silence. "Enough! Do you understand? I have said that *Njonja* Deibler is the new head of the camp." Turning on his heel, he left.

When he was out of earshot, I felt as if I were surrounded by a pack of jackals. "How could you?" "You're only a girl!" "We're older and more experienced!"

"Please, listen to me! I didn't know anything about this. I did *not* ask for the job! I do *not* want to take Mrs. Joustra's place, and I'm going to tell Mr. Yamaji so. Now!" I couldn't help but notice that none of them was protesting that Mrs. Joustra shouldn't be replaced!

I saw Mrs. Joustra leaving the building, so I ran to catch up with her. "Mrs. Joustra, I'm so sorry. You're such a good camp leader; I could never fill your place."

"Yes, *Mevrouw* Deibler, you can. You have been a good *loods leidster* [barracks leader], and I've talked with the commander about this. I *asked* to be relieved. *Mevrouw* Deibler, I'm so tired. I'll give you all the help I can, but you can do it. Mr. Yamaji has great respect for you, and holds you in high regard. He said that you were trustworthy: if you said yes, you meant yes. If you said no, you meant no. You may count on my support."

She looked so weary as she shook my hand that I could say no more. Hurrying back to the jungle camp, I prayed for wisdom. I wanted to tell Ruth, Lilian, and Margaret what had happened. I needed their advice. I found Ruth sitting in the doorway in the sun. "Ruth, listen! Mr. Yamaji has just announced that he has made me head of the camp!"

Before I could tell her what had happened, she said, "I've been expecting that."

"Ruth, you're no help. Why would you say that?"

"Because he respects you."

"But Ruth, some of the other barracks leaders are angry. They don't seem to understand that I had nothing to do with Mr. Yamaji's decision, nor did I know anything about it. I really don't want the job." Some of the others who had gathered said that Ruth was right. Didn't I know that after the Kempeitai had taken Margaret, Philoma, and me away, the commander would angrily jump on his bicycle and go out the back way to a village whenever the Kempeitai car was seen approaching? He wouldn't return until he was sure they had left. No, I didn't know that. Another lady, who worked around the office, told us that when the Kempeitai left a note for Mr. Yamaji, stating that *Njonja* Deibler was dying of tuberculosis and would never be returned to the camp, Mr. Yamaji left for Macassar at once. He spent three days going from office to office before he finally received permission to enter the prison and see me. The women—not only from my barracks but from other barracks and the Ambon camp as well—continued to gather and to encourage me to accept the position. I still had reservations in my own mind.

"Brace yourself: the opposition approaches!" someone said. It was indeed the leader of the ones who had verbally attacked me in the classroom. As the crowd continued to gather, Mrs. Heiden, red-faced, hair in stringy disarray, proceeded to tell the gathering how unsuited I was for the position. I was much too young. "Furthermore, this is a Dutch camp and we will *not* have an American to rule over us!" The eight Dutch families of my barracks were ready to cross swords with her. They were in my barracks at their own request, and we appreciated them.

The ridiculous side of the whole fiasco struck me as being very funny and it took some restraint not to call out to her, as she flung herself around and stomped off, "You'd better be careful, *Mevrouw,* or your sheaves will be bowing down to my sheaf!" One of the women pointed out what was all too apparent: it wasn't that I was too young, or that I was an American; Mrs. Heiden was jealous and wanted the position.

I don't know who told Mrs. Joustra about the woman's diatribe in front of the other women, but she came a short while later to say that she had heard about it and that she and the other woman had been friends for a long while. "But I told her that if she did not apologize to you publicly, before the end of the afternoon, we were no longer friends and I would never speak to her again. I will be back to find out if she has apologized to you!" I was very sad to think that a friendship might be broken over so small a matter.

About an hour later, the woman arrived, very humiliated and agitated, to say that she was sorry for the things she had said, that she knew I had had nothing to do with the commander's appointing me as Mrs. Joustra's successor, nor did it really matter that I was the youngest of the *loods leidsters* or that I was an American, since I spoke Dutch as well as Indonesian. I'm sure it was a very difficult thing to do, and I was glad she valued Mrs. Joustra's friendship that much.

After all the clamor was forgotten and things had returned to normal, I felt it was a propitious time to talk to Mr. Yamaji. I thanked him for the honor he had done me in appointing me as head of the camp. Then I suggested that, since the majority of women in the camp were Dutch, it might be well to have a Dutch person in the place of leadership. He asked if I had someone in mind. "Yes, I do—*Njonja* Bartstra of the Ambon camp. She is physically well, and a strong person, and I'm sure she would deal fairly with everyone, as *Njonja* Joustra has done."

"Have you talked to her?"

"No, *Tuan,* I wouldn't talk to her until I had asked your permission. If it's all right, I'll go speak to her now."

He was quiet, looking off into the distance as if deep in thought. Finally he answered, "All right, you talk to her to see if she agrees, but I'll need you later."

I ran to find Annie Bartstra, puzzling about the meaning of his last statement: "I'll need you later." Annie acquiesced, and I was grateful.

So much had happened in recent weeks. I opened the drawer of my memory and laid July 1945 on top of all the other months and years of my imprisonment. "Someday, Father, we'll look at it again, You and I. Until then, know that I have no complaints to make. There's so much I don't understand, but my heart tells me You do all things well." We moved into August, the eighth month of our fourth year of captivity.

The school was complete and in use. Repairs on the old sewing room were in progress. Sewing machines were being serviced, and work would probably begin there in a week or so. Undoubtedly the Japanese were in need of new uniforms. "That explains why they are setting up the sewing room again," we concluded.

Would the Japanese be replacing the barracks? No materials had been stockpiled as yet. Whatever they built would certainly be an improvement over the jungle camp, not to mention the added convenience of being closer to the wells and latrines, which now had scanty walls around them.

Late one afternoon, after filling hospital vats, I was making my way to the well to draw water for bathing when three ladies approached me. They looked about to make sure we weren't being observed; then one of them asked, *"Mevrouw* Deibler, who is Truman?"

"Truman? Truman who? I don't know anyone by that name."

"He's president of the United States," she whispered.

"Oh, him! He's our vice-president, Harry Truman, from Missouri. Our president is Franklin Delano Roosevelt."

"President Roosevelt is dead, and this Truman is now your president. Hadn't you heard?"

I was stunned by this news. "No, I hadn't heard. I knew Truman was vice-president but had never heard much about him. Thank you," I whispered and hurried on to the well. I didn't want to know who, or when, or where. What you don't know cannot be beaten out of you! The Brain's face loomed large in front of me. I could feel the pain of his crushing my arm and hear him saying, "If you ever have contact with anyone outside this camp, I'll get you! If you ever tell anyone about what happened to you, I'll get you the next time!"

As they turned to go, one of the women mouthed, "Pamphlet from outside!"

Terror made my pulse race. I hurried with my bath, all the while praying that God would protect me from any involvement with this pamphlet and protect all the women in the camp from the Kempeitai. Stepping outside, I looked all around; there was no one in sight. I walked quickly back to the jungle camp, quoting aloud the twenty-seventh Psalm.

When we were alone, I shared the news with Lilian, Margaret, and Ruth. Nothing more was said until it was apparent that it was common knowledge that someone had gotten a pamphlet from a source outside the camp.

Material arrived for the sewing room, but it was not material for Japanese uniforms, nor was it the course navy-blue material used for our work suits. Someone had taken notice of the threadbare condition of our clothes. We were to have dresses! Oh, that beautiful material of the most gorgeous pastel colors, soft blue, pink, yellow, green, and mauve. Mine was a lovely pink. We felt almost reluctant to wear the dresses, but we needed them. There certainly was no special occasion in the offing for which to save them. After the work was over, we bathed, then put on our dresses, but we changed back into our work suits for sleeping. We were careful with our dresses, but the time came when they had to be washed. The wash water turned white from the enormous amount of sizing in the material. Our lovely, crisp-looking dresses turned to pieces of cheesecloth—and the kind that definitely needed ironing. But the colors were still bright. Without slips, and in the sun, we looked like cadaverous spiders draped in sheer, colorful webs. With each washing the material became more sheer, so those of us without slips wore our dresses over our work suits.

Then came the shoes! Dump trucks arrived late one afternoon, spewing out white canvas tennis shoes into piles on the ground. "Pick a pair!" we were told as the trucks made a circle around us and left.

"Shoes! Shoes!" echoed and reechoed throughout the camp area. Women and children came running from every direction, and a free-for-all began. The shoes weren't tied in pairs, nor were they dumped according to size. If the Japanese had scrambled them as a prank, it backfired! They could never have guessed the fun and laughter the jumble brought to isolated, war-weary women and children!

It became a game. There were more than 1,600 women and children frantically pawing through 3,200 tennis shoes, each trying to find a pair that fit! Finding a single shoe that fit, we put it on and wore it. The game then was to find its mate. That was something else! One of the ladies finally settled for two shoes for the right foot. She put them on and asked, with a grin, if I thought they looked all right. "Well," I answered, "you'd better hang onto them. You'll probably never find another pair like them!" This started us giggling; she looked so ridiculous with two right feet. We became so weak from laughing that we had to sit down on the ground. Others joined in the infectious laughter. I think she kept her "laughing shoes"; they were always good for a laugh, whenever she put them on. I never found a mate for my right shoe, so I left it on the ground.

There were tugs-of-war, some name-calling, and a few real donnybrooks—two people wanting the same shoe. But all the confusion was a great release of anxiety, and most left the field of conflict the best of friends. Even if we didn't walk off in high-fashion footwear, it had been a wonderful afternoon.

Planes were still flying out toward the west mornings and returning in the afternoons, though no sound of bombings or antiaircraft fire could be detected. An increasing number of Japanese vehicles and visitors were in and out of camp. The visitors were all officers. We were intrigued. After the day's work was finished and we had eaten and bathed, some very anxious discussions took place as we tried to guess the significance of these activities as they related to us.

"Is it possible they are planning to move us elsewhere?"

"Could be. They are certainly doing nothing about new barracks."

"Would they be planning a new offensive?"

"That's possible. Do you remember last year? It was just like this, months of air raids—of course, they didn't bomb us, but the raids stopped, the planes left, and nothing more happened until three months ago. Well? As far as we know, there haven't been any recent raids."

"True, but . . . we don't even know whose planes those are that we hear. These visitors could be from some ships in harbor. Do you remember all those *hooge bezoeken* visits by Japanese officers?"

"No, I don't think they are *hooge bezoeken*. If they were, we'd be ordered out there to bow and scrape to them, while sprinkling the paths so they wouldn't get their boots dusty!"

"Is it possible that the war's over—well, almost over?"

"Don't get your hopes up. If it were nearing the end, we'd surely have had pamphlets to let us know, instead of bombs! And besides, they would have been allowing letters from and to our husbands. And they would have done something about our food situation."

"Well, what about the dresses and shoes? Don't you think that was to make us more presentable when the Allies arrive?"

This always capped the stack of suggestions! We would look down at our cheesecloth dresses, worn-out work suits, and mismatched tennis shoes—then a general, "Oh, definitely!" which always elicited laughter.

Having discussed in this manner all the pros and cons of our situation, we moved inside. How much we all wanted to believe that it would not be long before we could swing wide the gate to the outside world and run free again—free from fear of the Kempeitai, from sickness and separation from loved ones and death, but most of all, from confinement. The loss of our liberty, our freedom to contact the world beyond the barbed wire and the moat, that was the most grievous loss of all. Turning our thoughts to the One Who had sustained us thus far, we prayed together and found rest and courage to face the future.

August slipped away. Then the long-awaited day came. Most of us were thinking, when asked to assemble in the large open field where the memorial service had taken place, "They're going to tell us we're going to be moved." Commander Yamaji and the second-in-command appeared in full-dress uniform. Mr. Yamaji informed us that His Imperial Highness, the Emperor Hirohito, had announced by radio that the war was over and Japan had accepted the terms of the Potsdam Declaration for unconditional surrender!

Mr. Yamaji had been in conference with the Australians in Macassar as this area would be under the Australian Army of Occupation. Arrangements were being made for the internees of Kampili to be evacuated to Macassar, as soon as housing became available. Wives and families whose husbands were in Pare Pare would be given first priority. The rest would follow as soon as possible. They asked our cooperation so that the evacuation could be expedited in an orderly manner. Mr. Yamaji thanked us, saluted, turned, and left.

I have seen photos of the wild victory celebrations that took place in New York, San Francisco, and similar places when the announcement of VJ Day was broadcast to the nation—crowds singing, dancing, drinking, and kissing (whomever!) in the broad, confetti-filled, brightly lit streets!

It wasn't like that in Kampili. We were not safe on home soil, nor outside the barbed wire, nor half a world away from the battlefields, some still wet with the blood of fathers or sons or brothers. We were still within our prison confines, still separated from our families. We had nearly four years behind us of *total isolation from the rest of the world,* wondering how that world had changed, and who of our loved ones would be left. There was not even a conquering soldier in sight who had come to set us free, whom we could thank, whose hands we could kiss and wet with our tears of gladness. The full import of what we had just heard would come later.

It was a silent celebration of tears rolling down gaunt faces burned deeply while laboring in the sun on roads, in rice fields, in pig pens, on coolie lines loading and unloading trucks, emptying septic tanks; faces on which sorrow and suffering had etched their deep lines. For ours had been a silent war of waiting, and we had measured courage in simple endurance!

There was no riotous drinking. These were a people who had drunk deeply at the bitter waters of Marah, from the cup of isolation, separation, and the loss of loved ones—a people whose thirsty souls had just savored the first few cool, refreshing drops of freedom!

There were no happy songs filling the air, interspersed with shouts of victory. But I think the hosts of heaven must have hushed to hear the anthem of praise to our God and King, ascending from a thousand hearts or more, as our lips whispered, "Thank You, Father. From the depths of our being, we thank You!"

There was no wild cavorting about, just a quiet moving to and fro among the people to clasp hands, to embrace, to whisper thank-you's to those with whom friendship had been pledged in a mingling of blood, sweat, and many tears. Some drew back with eyes full of dread. They were the collaborators and the ones who had been unfaithful to their husbands. I had pity for them, but mostly I grieved for the children. Divorces and broken homes were inevitable.

Perhaps it is just as well that the scene was never photographed. Few could understand or appreciate the beauty or poignancy of our victory celebration.

Wandering back toward the jungle camp, I thought of those with whom I had shared the same barracks for the past few years in such close contact. I thought of Ruth and Philoma, Margaret and Lilian. I had come to have a deep love and appreciation for all of them. In those years of the very closest of associations, there had never been a quarrel or even cross words among us. I was the youngest, and I had needed their counsel, which they freely gave. They had loved me, encouraged me, and supported me in every way. These women of God truly adorned the Gospel of our Lord Jesus Christ. I am blessed to have known them. "How good and pleasant it is for brethren—and sisters—to dwell together in unity" (Psalm 133:1).

How kind, sympathetic, and supportive the people of Barracks 8 had been. They were a breed apart! There were many in the camp whom I considered to be true friends. I greatly appreciated the Salvation Army people, Father Bell, Mother Superior, and the sisters, and others who had prayed for Margaret, Philoma, and me when we were in the hands of the Kempeitai. Had it not been for these years, I might never have known these dear people, from whom I was soon to be parted. The French have a saying, *Partir c'est de mourir un peu,* "To part is to die a little."

The climax of that wonderful day was a Celebration of Worship and Praise! As a family, we worshiped and gave thanks to God for His care of us through the difficult times, now behind us. We commended one another to His continued care for all the years that lay ahead, expressing gratitude to God for each other, and praying for an abundant measure of strength and grace to be given those who would be leaving loved ones behind.

Emotionally exhausted, we slept.

Hurrying through the necessary tasks of bringing in water and vegetables, and food preparation, we went to the old camp to observe what was happening, hoping for word from Pare Pare. A car drove in shortly after we arrived. Excitement ran high when we saw a European among the Japanese. No one recognized him, so we figured he must be from the Australian Army of Occupation. We returned to collect our food and eat at the jungle camp.

Early in the afternoon, word came that I was wanted at camp headquarters. In fear and trembling, I ran to the office. Why were they calling for me? When I walked in, all those in the room stood to their feet—even the Japanese officers—then Mr. Yamaji introduced me to them. The Japanese clicked their heels to attention and bowed (that was a change!); then the European stepped forward to shake hands. He was an Australian major. Mr. Yamaji said that their Japanese translator was having a very difficult time. Would I translate for them? (So that was what he meant when he said, "I will need you later.") There were many important details to arrange regarding the transfer of authority and control from the Japanese to the Australians and Dutch. The classification and imprisonment of Japanese POWs was discussed. That was interesting; the Kempeitai were war criminals and as such would be under heavy guard in the prison in which they had committed so many atrocities. Adequate housing for the Kampili and Pare Pare people was vital. All the doctors, nurses, and patients were to be transferred to Macassar hospital immediately. The Lord gave me ready facility in the use of the languages for translating.

It was very gratifying to talk with the Australian major, who was the highest-ranking Australian officer in the British and American Military POW camp. He answered my questions about world affairs and described briefly what had happened during the last four years. The war in Europe had been over since May 7. MacArthur had made his first landing in the Philippines in January, and by March 4 some five thousand American prisoners had been released in Manila. The Philippines were now completely free. Most of the internees had long since gone back to the States. Japan had been severely bombed, and many landings had been made on the main island of Japan by the Americans. I listened carefully, trying to retain all he was saying to share with the other women later. He suggested that I might want to write my family and he would see that the letter was mailed.

It was of special interest to me that he knew Ruth's husband and had known Russell. He remembered Russell's ministry in his camp. For a time Russell was allowed to go to the British and American POW camp. He was able to help some of the men, who at that time were in very great need. These men were survivors of the Battle of the Java Sea and were routinely subjected to beatings with iron pipes, often as many as two hundred blows per beating—enough to reduce human flesh to pulp. He expressed sympathy, as Mr. Yamaji had told him of Russell's death.

I also learned that the planes we had been hearing were American planes evacuating the American and Australian POWs from the British and American camp, who understandably required medical help immediately. He understood that no women or children were to be moved by plane—but he thought it might be arranged. I told him of the other Americans in our camp—Margaret Kemp and Philoma Seely, and Mrs. Whetzel and her little girl in the Ambon camp. I mentioned that I did not wish to leave until I had had the opportunity of visiting Russell's grave. Mr. Yamaji joined us and said that the major's car was ready, so he left.

Joining the women, I shared all I had heard and observed. We had so much catching up to do. After devotions we talked far into the night. Excitement and anticipation welled within us. I made mental notes of all the things the women wanted me to ask the major if I saw him again.

The following morning I was again called to headquarters to interpret. I handed my letter home to the major. I had been very careful about what I wrote, lest my parents become unduly distressed. When the morning session was over, I came out to find the exodus had begun and Ruth had been in one of the early groups to leave.

An army truck arrived, but I didn't see the passenger. Someone reported that it was a Dutch doctor who had come from Pare Pare. I too was anxious to hear the news from Pare Pare. A woman touched my arm and said that Margaret Jaffray wanted me. She walked quickly away, and I thought she was off to garner more news. I went down to find Margaret and her mother in the Ambon camp. They were in one of the stone houses, and when I stepped in the door, I could see that both of them had been crying. A great lump of fear came up in my throat. "Margaret, what is it? It's not your father, is it?"

She dropped her head on my shoulder. "Yes, we just got word from a Dutch doctor that Daddy died last month, on the 29th of July."

"Oh Margaret, no! It can't be. I've been wanting so badly to talk to him about Russell's death. How terrible for you and your mother. Mrs. Jaffray, I'm so sorry!" I put my arm around her and we cried until there were no more tears. Mr. Jaffray was like a father to me. He was my mentor. I had been welcomed into their home, and they had all been so kind to me. I could not count the hours I had sat with him and Margaret while he shared things from his childhood, his home, his conversion, and his call to be a missionary. How dreadful for Mrs. Jaffray. He had always been there for her. What would Margaret do? Her father had been her very life. What would they both do now?

I asked if Lilian and Margaret Kemp knew. They did and had gone to collect their things as they would be going with Margaret and her mother to Macassar that afternoon. I was very glad that they would be in a different environment and that Lilian and Margaret Kemp would be there to help them. We had heard that most of the houses in town were disaster areas—filthy dirty, and most of the light and bathroom fixtures gone or broken. Many of the homes were without furniture.

Someone called that I was wanted at the commander's office. I kissed them goodbye and said I'd see them when I got to Macassar.

My heart was very heavy; I had to be alone for a moment. I walked to a nearby spring, knelt down, and splashed water on my face trying to staunch my tears. "Lord, this is a very bitter thing for Mrs. Jaffray and Margaret to bear. If only Dr. Jaffray could have been spared until they had some time together, or if they had been told a month ago, they would have had time to adjust, at least in a measure. They were expecting to see him this very afternoon. That is such an unwarranted cruelty on the part of the Japanese not to inform them at once. Lord, comfort them! And me: Lord, how much I wanted to talk to him about Russell's death. He would have known all about it. Did You really have to take him at this time?"

By the time I got to the office, the Lord had calmed my heart. A Dutch gentleman was there, who stood up when I entered. He looked inquiringly at me, then asked, "Are you Mrs. Deibler?"

"Yes, I am."

"What's your name?"

"I'm Mrs. Deibler."

"Sorry, I mean your . . . er . . . uh—" he paused, and I realized he was searching for a word.

"Oh, you mean my given name? It's Darlene."

"Darlene, Darlene, yes, you're the one. I'm Dr. Goedbloed. I'm the one who attended your husband when he was sick, and I was with him when he died. Please sit down." He proceeded to tell me that Russell had had dysentery, as had many others in the camp; but when their rations were cut and the food was poor, Russell lost a lot of weight and became increasingly weaker. "The problem was that our rice was musty and full of worms. If he had been well, he might have been able to eat it anyway, like the rest of us, but being so ill, he just couldn't. He tried, but it came back up. We made a rice gruel for him, but even that he could not keep down for long, so he became dehydrated. If it had just been the dysentery, I think I might have saved him, but he had a serious heart condition. That was the cause of his death. At some time he must have put a terrible strain on his heart, damaging it, and it just gave out. I'm so sorry. I did everything for him that I knew to do, but we did not have the medicines or the facilities we needed.

"I came because I wanted you to know that for the last four hours before he died, he kept calling for you, 'Darlene.' I and every man in the camp who we thought might have some influence with the Japanese begged them to fetch you, but they refused. He passed away at midnight on the 28th of August 1943. I wanted you to know that his last thoughts before going to heaven were of you, Darlene. He was such a good man."

I looked up through my tears, to see his eyes searching my face. "What a kind, thoughtful man you are, Dr. Goedbloed. I know that you and the others did everything you could to save Russell. Father Bell told me as much. You cannot know how much this means to me, what you have told me. I'm very grateful to you. I want you to know, too, how much we have appreciated your wife here. She's done everything for us she could with very limited medicine and facilities, as have the other doctors. Your wife is a very kind and sympathetic lady."

He stood up to leave, and I thanked him again. As we shook hands, Sweet Seventeen came out of his room to shake our hands and bow to us. I knew that he, too, was sorry about Russell's death. This he had told me the day I was informed that Russell had died. Remembering what Dr. Goedbloed had just told me, I realized that Sweet Seventeen had been there in Pare Pare at the time, but he would have had no authority to arrange for me to be brought from Kampili if the commanding officer had refused, nor would he have had the authority to tell me about Russell's death when he arrived in our camp.*

I went to the well to bathe before returning to the jungle camp. I washed myself and my work suit at one and the same time, without soap. A good soaking couldn't hurt either of us. I walked slowly to let the sun dry my clothes, and arriving at the shack, sat on the stoop in the sun to commune with my Lord.

I heard a man's voice, and, looking up, saw a man in white clothes talking with some women. They all turned and looked in my direction; then one of the women pointed at me and waved. The man left them and walked toward me. He was wearing an immaculate white uniform and highly polished black shoes. He looked so squeaky clean and his hair was so neatly combed that I felt like an unkempt, bedraggled, skinny waif from the other side of the tracks. I hadn't seen so much "altogetherness" in years! I quickly ran my fingers through my hair, sat up straight, and smoothed my work suit.

His first question was, "Are you an American?" I knew he was thinking that I looked like no American girl he had ever seen.

I stood up as tall as I could and rather defensively answered, "Yes, sir. I'm an American!"

His eyes were locked on my bare feet. "Don't you have any

---

*Mr. Presswood later told me what a cruel, sadistic man their commander was, forcing them to work until they dropped with exhaustion at work for which there was no rhyme or reason. He withheld food from them, so the men had to scrounge for anything edible—even leaves from the jungle trees. When they were given food again, the daily rations had been reduced to 280 calories. They were forced on a death march into the far interior of Celebes after their camp had been bombed. For some time they lived in pig sties. They were flooded out, then traveled through four-to-six-inch-deep mud, all the while enduring a reign of terror by the guards, who beat the men unconscious for the slightest offense, then revived them by throwing water on them so they could beat them again. Dysentery was epidemic and more than twenty-five men died. Many others had beriberi, scurvy, and malaria. Dr. Jaffray kept busy writing Bible expositions in Chinese for the *Bible Magazine* in a notebook. He remained cheerful, always encouraging the others. When their rations were cut so drastically, he weakened rapidly, and on July 29, 1945, he slipped quietly and triumphantly into the presence of the God he loved implicitly. Faithful soldier that he had been, his final battle won, he laid down his sword and took up his crown of everlasting life . . . this just a month earlier.

shoes?" I slumped back onto the stoop and pulled my feet as far back under the shack as I could; then I became aware of the empathy in his eyes.

"No, but that's all right; I'm used to going barefoot. They recently brought in some tennis shoes, but I couldn't find a pair."

"Well, I'm going to get you a pair of shoes. What size shoes do you wear?"

"Probably a size 5 or larger now. But don't worry. I don't mind going barefoot." I realized that his abruptness was anger against those responsible for my condition, that he was feeling sorry for me.

"I'll send some shoes tomorrow. What else can I get you?" It was obvious that I needed clothes.

"Please, I would like so much to have a comb, and maybe some soap? We were bombed and I lost everything."

"I'm sorry. I should have introduced myself," he said. "I'm Tom Sawyer from Los Angeles, and I'm with the U.S. Navy."

It was on the tip of my tongue to say, "Yes, and I'm Becky Thatcher," but I didn't know if he had a sense of humor. I settled for, "I'm Darlene Deibler from Boone, Iowa."

"Look, the major told me about you, and I've been thinking: tomorrow, when we fly in, we could make a drop of some things to you. Do you know of a place here that is fairly level and free of trees? We'll be coming in low, as it will be a free drop."

"Of course! The rice paddy, right down there." We walked toward it. "Please, we need food for the children more than anything—powered milk, *any* kind of milk, cereals, flour, maybe some meat if you have it, and combs."

He thought the rice paddy would be excellent and suggested that it should be marked in some way, so the perimeter would be clearly defined. I told him I was sure I could get something from Mr. Yamaji for marking the drop area.

"I must go now—there's a car waiting for me—but we'll be back tomorrow."

Women appeared from every direction with a dozen questions. "Who was he?" "What did he want?" "What did he say?" "Why were you looking at the rice paddy?"

I answered all their questions, and they laughed with me about the Tom Sawyer bit. But when I told them about the food drop, they hugged and kissed me and we danced around singing. "Food! Food! Glorious food!" It didn't take much to start a revelry at this juncture—just a little *food*! Before it was dark, everything was organized. Mr. Yamaji very obligingly brought out a bolt of white material that we tore into strips for markers. Sweet Seventeen offered his help.

The drop went off without a hitch, except for poor little Broertje Routs's reaction. Even though the planes were flying very low, the impact of the drop caused the friction lids to fly off the tins, and the contents sprayed upward like geysers—except for one tin. There it sat, its lid still in place. When Broertje saw his mother start toward it, he jumped up and started screaming frantically, "Mother, don't touch it! Don't touch it—it hasn't exploded yet!" The tears were running down his cheeks.

I ran to pick him up, thinking he was going to faint, so great was his terror. I held him tight, repeating, "It's all right, Broertje. It's not a bomb!" I motioned to Mrs. Routs and the women to leave it. His older brother brought a tin and a lid to show him what happened when a tin hit the ground. When he was assured it wasn't a bomb, we fetched the tin, opened it, and showed him what was inside. Funny? No. We didn't laugh; I felt sick with hurt and anger for Broertje and the other little ones whose emotions had been ravaged by the sights and sounds of war.

During the gathering of the food that had been dropped, I came across a tin of sweetened condensed milk. The tin had burst, and the contents were dribbling away onto the ground. "What a terrible waste!" I grabbed up the half-empty tin, poured the milk into my hand, and lapped it up until the tin was empty. I licked my fingers and lips, still savoring that delicious, out-of-this-world flavor. My hands were sticky, so I set off to get some water when suddenly—I dashed behind the nearest tree! Wiping my mouth with a leaf, I philosophized that it was worth it, for it was probably the only time I would ever enjoy food going both ways!

We had been so hungry for meat, and at last we were to have some. The U.S. Navy had sent us some tins of meat called Spam, and we decided that Spam was "beast in its most excellent, succulent form"! The generosity of the navy in sharing so much Spam with us was quite overwhelming. (When later I mentioned to some of the men how liberal and kind they had been to us, they looked embarrassed, then burst out chuckling. "My dear, you are so welcome! Would you like some more?")

True to his word, Tom Sawyer had included combs and soap. Someone mentioned an additional package at the office with my name on it, so I went to see what it was.

I saw Mr. Yamaji standing in the doorway. After greeting me, he went back into the office and came out carrying a package designated "for Darlene Deibler, the American." He called something in Japanese to someone inside. I deduced that he had said, "Bring a pair of scissors," for seconds later Sweet Seventeen handed me a pair. Mr. Yamaji wanted me to open it, so I did. It contained a pair of shiny, black men's shoes. They were small, but not small enough.

Tjoba, Njonja. "Try them, Mrs." I put them on, and both Mr. Yamaji and his second-in-command thought they were very beautiful. *Bagus, Njonja!* "Beautiful, Mrs.!" I smiled on standing up, for I could see so much shoe out in front of me that I thought I must look like Sweet Seventeen with his very long feet. The shoes were too big, but I knew a woman whom they might fit and who needed shoes badly. Mr. Tom Sawyer was a very gracious man to send me the shoes. I hoped I would see him again to thank him.

Mr. Yamaji asked if I were going away with the other Americans. "No," I said, "I would like to visit Mr. Deibler's grave before I leave."

"Ah! *Tuan* major told me that, *Njonja,* so I have ordered a truck to be here tomorrow morning to take you to Pare Pare, and I have also sent word to the men there that you are coming. The truck will then take you to Macassar so you will be ready to go with the last planeload."

I was amazed that he knew all about the plane schedules and so forth. I thanked him profusely for arranging transport for me.

"It's nothing, *Njonja.*"

"Tuan Yamaji, I have wanted to thank you ever since you came to Macassar to visit me in prison. I was very sick from malaria, beriberi, and dysentery. I never had tuberculosis as the Kempeitai told you, but I was very sick and the bananas you sent me by the guard gave me strength so that I didn't die. Did you know that the Kempeitai men said they would cut off my head?"

He nodded. *"Njonja,* I told them you were not a spy. I knew you weren't, because of what you had told me about God, here in my office. I told them what you had told me. I'm sorry, *Njonja,* that your husband died and you have had so much trouble, but now you go back to America."

"Mr. Yamaji, I have told *Tuan* major about your visit and the bananas that saved me from starvation, also about the Kempeitai. I will tell *Njonja* Presswood's husband also." (Mr. Yamaji had pleaded mercy for me; I certainly owed him that much.)

Suddenly something occurred to me. Mr. Yamaji had told them I had talked to him about God's love. That explained why, at the last hearing, the Interrogator had said to me, "But if we win the war, you would not stay and tell my people about God, would you? You'd go back to America?"

"No, I wouldn't go back, if God wanted me here. Do you know that God loves your people too?" But the Brain spoke sharply to him and he said no more.

The Australian major from the Army of Occupation told me that the two Kempeitai officers, the Brain and the Interrogator, had cut their wrists on barbed wire in an attempt to commit suicide rather than come to trial. But they had been unsuccessful, and both were sentenced to be executed.

I said goodnight to Mr. Yamaji and Sweet Seventeen and thanked them again for arranging the truck for me. It was still light, and most of the women were in the area of the old camp. A vehicle had stopped outside the gate and the passengers were coming in. I saw Ruth and Ernie among them, so I ran to meet them. Ruth was absolutely radiant; the transformation was amazing. I felt happy to see them together again. We spoke briefly, as the car was not staying long. They were aware of my plans for the next day and wanted to be sure I stayed with them when the truck brought me from Pare Pare to Macassar. "Wiesje knows where we're living."

Ernie handed me four letters and a drawing. "There's a letter I wrote you after Russ died, but I wanted to tell you personally." His face was drawn, and he said softly, "Be thankful that the Japanese never fetched you. You would never have recognized Russ. Now you'll be able to remember him as he was the last time you saw him. I don't know why God took him. His life and messages had a great impact on the men in all the camps. I miss Russ; he was my dearest friend, and I think Dr. Jaffray would have said the same." He looked at me sadly, remembering. "May He comfort you. We have to believe that our Father knows best."

The driver sounded the horn, so they had to go. They assured me that they would be looking for me the next day. I waved, but I couldn't speak. Clutched tightly in my hand were the letters Ernie had brought.

I sought a place near our hut in the jungle camp where I had privacy to read my letters. The people understood; they had seen Ernie hand them to me. I looked long at a drawing by Brother Geroldus. It was a sketch of Russell's grave, beautifully done in pencil. At the bottom, he had written *God kiest de besten,* "God chooses the best." On the reverse side I found these words: "Drawn by Brother Geroldus, who was with him when he went to heaven, and who closed his eyes, on August 28, 1943." I prayed that God would bless Brother Geroldus for his kindness and the Christian love he had shown Russell and now me. I laid the drawing carefully on the grass beside me.

Then, with trembling hands, I took up a letter carefully folded to make an envelope. I recognized the writing. It was addressed to Mrs. C. Russell Deibler, Kampili. The letter had been written not long before Russell died. "My dearest darling," the letter began, followed by what he dared to write about conditions. His expressions of love again included the sentence that I will never forget: "My darling, I have wished 1001 x's that I had taken you away from here. I am concerned for your safety." Commending me to God's care, he closed with, "I send you all my love, Russell."

It took me a long time to finish the letter. I didn't want to stain it with tears. I was so grieved that he felt he should have taken me away. Both of us had agreed that we should remain, and that decision was reached only after much prayer. "Lord, I trust that You reminded him that it was You Who impressed upon both our hearts that we should *not* leave. I have been safer here, overshadowed by Your love, than I would have been anywhere else on this earth, outside of Your will!"

The first letter from Dr. Jaffray said, "I owe Darlene Deibler $25.00 for her twenty-fifth birthday. Payable after the war. Happy Birthday, Lassie! Dr. Robert A. Jaffray." I loved it that he had been thinking of my birthday and no doubt praying for me.

His other letter and Ernie's were both written after Russell's death. The contents were similar to what Dr. Goedbloed had told me, except that Dr. Jaffray made mention of Russell's asking to be anointed with oil and prayed for, which he and Ernie had done.

"That was around 11 P.M., after which Russell seemed more restful. I do not remember having prayed for anyone like I prayed for him, but somehow in the providence of God it was not His will to heal him. It is for you, Lassie, that I pray, that you may be comforted and accept Russell's death as the will of God."

Ernie added that "Dr. Jaffray and I had gone outside. Hearing Russ call for you—we couldn't bear it. We walked back and forth outside the room where the doctor was attending Russ. We prayed that God would spare him, not only for your sake, but for the sake of the field. He was my dearest friend and God wonderfully used Russ from the time we were taken to Macassar, in our camp as well as the British and American Military POW camp; the men there were in great need. His messages left such an impression on the men that months later, they could still relate the main points. Russ had wonderful rapport with men of all classes. His death brought a feeling of great loss, felt throughout the camp. What Brother Geroldus wrote under the sketch of Russ's grave, *God kiest de besten,* expresses what we all felt. God had indeed taken one of the best men in the camp to be with Himself. His funeral was very impressive and we were all deeply moved. Many of us wept openly and unashamed. Some months later a Dutch gentleman told me that at Russ's funeral service, he had given his heart to Christ. I miss Russ, and there are many others who sorrow with you."

When I went into the hut, the women who were left were waiting. I shared Russell's and the others' letters with them and showed the drawing by Brother Geroldus. Our lives had become so interwoven that what mattered to me mattered to them. We had become a close-knit family. By the time I finished reading the letters, we had drawn into a tight circle, each drawing strength from the other. Clasping hands, we prayed. I saw the measure of their love for me and mine for them expressed in tears, knowing that after tomorrow, our paths might never cross again.

After everything was quiet, I crept into the throne room of my heart. I had not come as a petulant child to assert that a privilege had been denied me, but that I had a need to confess how wrong I had been in misjudging the whole situation. I had felt grief and anger that Dr. Jaffray was dead and I was denied the privilege of hearing from him all about Russell's death, and frustration at my need just to talk to him. "Forgive me, Lord, for not ten minutes later, I was sitting in the commander's office with Dr. Goedbloed, who could tell me all about Russell's sickness and death. Did anyone else know about Russell's damaged heart and that if he had lived, he might have been an invalid? Russell, an invalid? What a terrible blow that would have been to a man who had been a pioneer all his missionary life. O God, I must believe You do all things well!"

The doctor had asked me if I knew when Russell's heart might have suffered damage. I remembered the day in Macassar when he came down the gangplank of the ship returning from his first trip over the trail to open the New Guinea field. He had made the trip alone; he was denied sufficient sleep by carriers plotting to drop their loads and abandon him; he suffered a too-rapid loss of weight due to insufficient supplies; he endured the rigors of a trail—that *was* no trail—following a ribbon of water to find a Stone Age people "who had never heard," and to claim, by faith, a land and its people for God that New Year's Day 1939, to plant a seed and water it with his life's blood, that from New Guinea's uncounted multitude there might be the redeemed to sing with us, "Worthy is the Lamb who was slain."

Then came the letter from out of the past, Russell's letter, preserved by Ernie, who understood what it would mean to me; kept since 1943, affirming the love the doctor surmised. Pure love always leaves behind a legacy of undefiled memories without regrets.

It was wrong of me to think I was being denied a privilege. Written as he would speak, Dr. Jaffray's letter shared with me all about Russell's death, with the added dimension of the anointing of Russell. God did heal him, then took him home—his heavenly home, where there is no more sickness, no damaged hearts, and no more pain. It was lovely, too, to know that even though miles and a war had separated us, Dr. Jaffray had remembered my birthday and had prayed for me.

Ernie could have destroyed his letter when peace came, but he didn't. The hurt from the loss of his friend, Russ, lay fresh upon the page. How encouraging to know that God had graciously used Russell's ministry to comfort, challenge, yes, and convict others. "He being dead, yet spoke," and at his funeral, a man had given his heart to Christ. Ernie, too, had told me all about Russell's death and burial.

Recalling the drawing by Brother Geroldus and the words, *God kiest de besten,* it spoke comforting words that Russell's life had adorned the Gospel of Jesus Christ, and he had been loved by those who knew him. Brother Geroldus could not have known how beautiful that drawing was to me.

I prayed the Lord to forgive me; I had asked to hear from one person and my Lord had given me five responses to that request. Once again my Lord had done the "exceeding abundant" above what I had asked. That's a ratio of five to one. I was reminded of another time when it was ninety-two to one! How good my Lord is!

I rose early the next morning, to bathe with soap and comb my hair! Then I picked up Mrs. Lie's Bible, my letters and the drawing, my comb and soap, my spoons and the burned trinkets, rescued from the ashes. Piling them neatly on a rag, I tied a hobo knot to facilitate carrying them and walked away. After a few steps, I paused to look back at the old shack. Within its walls, we had experienced the heights and depths of emotions. Our nerves had been stretched almost to the breaking point. We did not write the scenario; we were but the reluctant actors. "But the drama of these past two months will be remembered long after you, Old Shack, have fallen to your knees, rotted, and been reclaimed by the jungle floor!"

I joined the others drifting toward the headquarters. When the truck arrived, to my surprise and delight Wiesje Kandou stepped down to greet me. "O Wiesje," I squealed, "I had no idea you would be coming." She handed me a sack, saying that Mr. Yamaji had said she should bring me a dress and shoes! So he had thought the other shoes, though they had a beautiful shine, were not appropriate with a dress. How thoughtful.

I ran to the well enclosure to change as quickly as possible. "Bless you, Wiesje!" She had included some lingerie. The material of the dress I recognized immediately. I knew its history. Helen Sande and her husband, Andrew, were fellow students in St. Paul and long-time friends. Baby David Jerome was born in Macassar before they left for Borneo. Helen had asked me to buy material for two dresses and have Tante Mien make them like a dress of mine that she had always admired. Wiesje's aunt had evidently not been able to get them sent because of the war. I was sure that Helen wouldn't mind my borrowing this one. It was made of a blue rayon material with splashes of black and white for design and a black-and-white tie belt. I slipped the dress over my head and looked down to see that it nearly reached my ankles. "Never mind," I thought, "I'll cinch the belt in very tight and blouse the top over it." It didn't look too bad. The shoes were white sandals with ankle straps and high heels. They fit quite well, but I needed a bit of practice walking in heels again, so I circled the well several times. I would find it very embarrassing if I fell off my shoes in front of the other people. I got out my comb, ran it through my hair several times, then slipped it back into my "rag-bag." Did I look all right? I don't know, but "look out, fashion world, here I come!"

Stepping out of the enclosure, I postured, then walked with mincing steps, holding my rag-bag as though it were a fashionable bag. Cheers went up, punctuated with clapping and calls of, *"Mev-rouw* Deibler, you look beautiful!" Never forget, beauty is in the eye of the beholder.

I had to make a joke of it to keep the lump that was rising in my throat from getting the best of me. These were friends to whom I was saying goodbye. Most of them would be going to Macassar today, so we parted saying, "I'll see you in Macassar!" knowing, in reality, that we might never see one another again.

I went over to Mr. Yamaji and Sweet Seventeen, who stood talking to the Japanese officer who had come on the truck. Mr. Yamaji introduced us, mentioning that he was like me. I didn't understand what he meant, so made no response except to smile and nod. He put out his hand and we shook hands.

I looked toward Mr. Yamaji and Sweet Seventeen, "I go now, *Tuan Tuan* "gentlemen." I want to thank you for your help in getting the truck for me. May God bless you both. Remain safely, *Tuan Tuan."*

"Thank you, *Njonja.* A safe journey." We bowed to each other, then Mr. Yamaji went into his office and closed the door. That was the last time I ever saw him.

Sweet Seventeen watched till Wiesje and I were seated in the cab of the truck and the Japanese officer stood in the bed of the truck. We waved and called goodbye to everyone. Sweet Seventeen raised his hand in a salute, probably to the man in the back.

Wiesje and I talked nonstop, catching up on the news of her family and the friends from the Tabernacle. It occurred to me to wonder how Wiesje knew Mr. Yamaji, since she said he had asked for the dress and shoes for me. "I don't know him, but he told the chaplain to get them and the chaplain came to me. When he told me they were for you, I told him I was coming along as you were my friend."

"Wiesje, who's the Japanese in the back? Mr. Yamaji introduced us and said he was like me."

"That's the Japanese chaplain. I thought you knew. He really is a true believer."

The truck slowed, then turned into what had been the Pare Pare Dutch police barracks. A gentleman stepped forward to help us down from the truck; then others came over to introduce themselves and shake hands. Some I recognized from Macassar. The first gentleman said that they had morning coffee ready, if we would like some; then we would go to the cemetery. Was I agreeable?

I thanked him for making the arrangements. Celebes highlands coffee is famous for its flavor, and they had the real thing to serve. It tasted wonderful after the many-times-used, reroasted coffee grounds we had been using. I asked if I might have a second cup, which pleased them. We were speaking both Dutch and English. Some of the men were busy in the kitchen.

The cemetery was not too distant from the camp. The men had cleaned around the graves, and each grave had flowers on it. Someone pointed to a nearby grave and said, "This is Mr. Deibler's grave." I looked at the wooden cross. There was no name on it, so I decided that Brother Geroldus must have just sketched a cross on a grave; his was not intended to be a true reproduction. I knelt, thinking, "It's like an unmarked grave." I could hear a discussion going on behind me; then a hand was laid on my shoulder. "We're very sorry, but the man made a mistake. This is not Mr. Deibler's grave. His is over here."

"Please don't feel embarrassed," I said. "It's all right." Whoever had made Russell's cross was a master craftsman. It looked like the drawing. The flower arrangements on the grave had been done with care. Everything was an expression of love for the man whom they had buried here two years earlier. I wept softly as a terrible sense of aloneness swept over me. "When I leave here, Russell's grave will be so very far away." Then the Lord gently reminded me, "You do not sorrow, as those who have no hope

(1 Thessalonians 4:13). Here are but the remains of his earthly body; he is at home with Me. My child, this is not where you belong; there is a world of lonely, hurting people out there. Some are right there behind you!" With the hobo knot of my rag-bag, I dried my face and turned to join the waiting men. They stood with bowed heads, remembering the man; and one, no doubt, was remembering that at this graveside he had invited Christ to come into his heart.

Back at the barracks, four long tables had been placed to form a square. The places were set around the outside of the tables so that everyone could see everyone else with ease. They had set places for Wiesje and me in the center of one of the tables. There before us was an enormous bowl of beautiful white rice, and other serving dishes piled high with food such as I had not seen in years. It had been cooked by the men in Indonesian style, my favorite cuisine. The realization of the sacrifice these men had made, and the labor expended by them, nearly took my appetite away—but not for long! I couldn't help it—I ate like a starved POW! Dinner over, the benches were pushed back a space from the tables. What followed may have been planned, but there was a spontaneity that made me think it hadn't been previously arranged.

A gentleman at one of the tables stood up, introduced himself, and told what Russell had meant to him. Then someone at another table, then another, always introducing themselves, spoke, sometimes recalling an amusing anecdote at which we laughed. Often the stories were very personal, but no one was embarrassed.

Certain phrases or words impressed me. In describing Russell's personality, people spoke of him as kind, patient, helpful, cheerful, and an inspiration. One man said, "He cared about us as individuals, and no matter where or when, he made himself available to us. He was a good listener and seemed to know when we just wanted someone to listen to us." Two men whom I had known before the war mentioned that when Russell was praying, they opened their eyes to look at him, half-expecting to see the Lord standing there. "It was as though he were talking to someone who was right there next to him." They made mention of the practical, timely messages that Russell gave. "His circumstances were our circumstances. He ate what we ate, he endured what we were experiencing, so he knew how to explain to us the source of his help, which was God."

There were few who did not make reference to the Christmas Eve service of 1942. The whole camp had gathered and Russell had related the story of "The Other Wiseman" so clearly and simply that even the Dutch men, who were not fluent speakers of English, could understand. The story had left a lasting impression on them. Father Bell had told me of this service, but even so I was moved to tears.

Wiesje came to my rescue when she pressed a large white handkerchief into my wet hands. I had been wiping my face with my hands, then my hands with the hobo knot of my rag-bag.

When I stood to respond, my heart was crying to God for the right words and the ability to speak without breaking down. I had been mulling over many thoughts in my mind. I had never realized that I would have the opportunity of telling them face to face how much their kindnesses to Russell had meant to me. This day that they had given me was a treasure. At the cemetery I had seen how much physical energy they had expended in cleaning around the graves and gathering flowers, when they should have been resting and recuperating after the terrible ordeal they had just experienced on the death march!

"Many of you had not met me prior to this occasion, and I recognize that all this was not for me per se. This has been like a testimonial dinner, a memorial service to honor a man you loved, a man whom God used to minister to you in your time of need. This was a very beautiful way of saying farewell and thank you to your friend once more before you leave this place. Thank you for allowing me to be a part of it.

"I've been thinking much lately about the balance in this matter of giving and receiving. Jesus spoke of it in these words, 'Give, and it shall be given unto you; good measure, pressed down, and shaken together, and running over' (Luke 6:38). Mr. Jaffray, Mr. Presswood, and Dr. Goedbloed mentioned how you have all ministered to Mr. Deibler; then we are still your debtors. I have mentioned this to my friends in Kampili, especially those of Barracks 8—wives of some of you here. I gave to them, but with what abundant, overflowing measure I have received from them again and again. I pray that God will return to you in abundant measure the many kindnesses you have shown me today."

I sat down and for a long moment no one stirred. Finally several men left the table and went outside. Others came to speak with me. Brother Geroldus introduced himself, and I told him how very much I appreciated the drawing. "In fact, I have it here, carefully wrapped around my Bible so I don't have to fold it. It's very beautiful, especially what you wrote on it." It pleased him that it was meaningful to me.

After the others went outside, a Mr. van der Haarst stepped up to hand me a notebook. I recognized it immediately as the one I had placed in the pillowcase for Russell when the Japanese had taken him away from Benteng Tinggi so long ago. He said he had been teaching Russell Dutch. We talked about his lovely family, who had asked to come into our barracks for the Christian fellowship. He knew of his daughter's death, as the women had been permitted to send a letter to their husbands shortly after Leintje passed away. He mentioned that there was a poem in the notebook for me. "I want you also to know that at the funeral, I opened my heart and asked Christ to come in. I know now that I shall see my dear friend Deibler again. I have been praying that God would comfort you, too." My heart was thrilled, thinking what a happy family reunion awaited him in Macassar.

The men who had gone out first came to say that the truck was now ready, but they wanted to ask about Mr. Yamaji. Did I know that just before he was transferred to Kampili, he had beaten a man to death? They had all been deeply distressed, thinking of a man like that being in charge of their wives and children. "But we hear he is a changed man. Is that true?" I told them of my contact with him after learning of Russell's death, and of how he had visited me in prison.

"Hmm, well, we said, 'It took the women to tame the tiger!' Mrs. Deibler, we have some fruit and vegetables we'd like to give you. These are eggs—maybe you'd better carry them by hand." I really didn't know what to say, for on the bed of the truck were three copra bags full of all kinds of things.

They stood waving to us until they were out of sight around the bend of the road. It was a day for remembering.

When we drove to Macassar, it was still light. Wiesje wondered if I would mind stopping on the beach for a while; the chaplain wanted to talk with me. I liked the suggestion, so Wiesje told the driver where to go near the sea wall, where we could sit down.

We talked about the war and the effect of peace on both our

countries. Then he made a statement: *"Njonja,* some people would not understand if I said this to them, but I think you will. I'm thanking God we lost the war." I looked up quickly to see if he realized what he had just said. "I really mean that! We are a proud people, and if we had won the war, the doors of Japan would never again have opened to missionaries. Many people would say I was a traitor to my country, but I love my country and my people enough to suffer the humiliation of defeat, that they might have the opportunity I have had of hearing that Christ is the Son of God, and that He died for all. There are many among the soldiers of my country who are now asking questions. There is a receptivity to my ministry, since the worship of our ancestors has failed to give us the victory. They are searching, and I ask that you pray for them. I pray for you, and I'm sorry about your husband. You will go back to America now?"

"Yes, my furlough is three years overdue. I want to see my family, and I need rest."

"Will you return to Macassar?"

"That I cannot say, but I *do* know that I will always be a missionary. I sometimes wondered if God might not be preparing me to be a missionary to your people. However, my Lord will lead me, and I will know where He wants me to minister." We talked mostly about spiritual things, and from his responses, I felt that he really loved the Lord. He said that he was an army chaplain because that was the will of God for his life. This explained Mr. Yamaji's statement to me: "He's like you." I also learned that he and Mr. Yamaji were friends and that they had talked about the Lord. My heart praised God because of this encouraging news.

Wiesje said that we had better go, so the chaplain waved for the driver to take us on to the Presswoods. He and the driver carried the copra bags of fruits and vegetables into the house; then they left to take Wiesje home, as it was getting dark.

I handed the eggs to Ruth, who said, "My, my! What have we here? Eggs? I'd almost forgotten what an egg looks like!"

Ernie helped us unpack the copra bags. There were so many hands of bananas, that when we laid them side by side, they stretched from one end of the fairly large lounge room to the other. "Well, Ruth, some people may have wall-to-wall carpeting, but we have wall-to-wall bananas!" I shared the events of the day with them as we started eating our way through the wall-to-wall bananas. I was scheduled to leave the next afternoon with Margaret Kemp and the Whetzel family. Miss Seely didn't wish to leave.

Borrowing a needle and thread from Ruth, I hemmed my dress to a proper length. In the bedroom Ruth had cleaned for my use, I found a single bed with a kapok mattress on a wood frame. That was real luxury, a marvelous improvement over the slatted floor of the old shack in the jungle. I lay down, closed my eyes, and two minutes later opened them to find that it was morning—at least, it *seemed* that way. I had been both physically and emotionally drained by the trip to Pare Pare.

Many came to wish me a safe trip back to America. Women brought husbands to meet me. I was glad to meet those for whom we had been praying, and to know that God had answered prayer. How lovely to see their happy faces and to know that, at long last, they could cast aside "the spirit of heaviness" and be clothed with "the garment of praise"!

Shortly after the noon hour, Wiesje arrived in the truck to say we had to leave immediately, as the truck was needed elsewhere. Mr. Yamaji had arranged with the Australian major to put the truck and driver at my disposal until I left on the plane. Wiesje said she would tell the chaplain to thank them for me. I said goodbye to Ruth and Ernie, never realizing that Ernie would outlive VJ Day by only a few months.*

Arriving at the harbor, we found Lien (our cook), her sister, Beth (the Jaffrays' cook), and Roda (the Jaffrays' laundryman and gardener) waiting for us. It was a sad meeting and a sadder parting. Lien had a handkerchief of Russell's, which she asked to keep for a remembrance. Beth had come to say goodbye and to share my grief in the loss of Russell and Dr. Jaffray. They had remained in Macassar so that they could help us "when the war was over." Roda had been crying with us, and when I asked about his wife, Rassing, he said that that was what he had come to tell me. She had remained in the mountains while he was looking for work in

---

*Ruth, while lying ill herself in a hospital, learned of his homegoing. Dear man of God, the last of the Macassar male staff, he had succumbed to the deprivation and cruelties of the imprisonment. It was a privilege to have known and worked with Ruth and Ernie in Macassar. An unselfish, kind, loyal friend, a servant of God and his fellow workers, greatly beloved—such a man was the Reverend W. Ernest Presswood.

Macassar. It was very difficult, because the Japanese paid only half of what he had received from the Jaffrays. Rassing had become very ill and they sent for Roda, but she died before he could get to her. "She left a message for me, to say she was going to be with Jesus and for me to believe in Him and get ready so we could meet in heaven. She wanted you to know. I've waited for you to come to Macassar so I could tell you about her and that I'm ready too!" How well I remembered her little brown pixie face, aglow with the joy of the Lord, that morning four years earlier when she gave her heart to Jesus Christ! And now the joy of knowing that Roda, her husband, was a believer, too! Out of death came life— eternal life for a dear Boegis man!

The Whetzels arrived as the Catalina was taxiing in toward the buoy; a dinghy took them out to the plane. Then Margaret Kemp arrived. As we stood waiting for the dinghy to return, I began closing the door on this epoch of my life. I turned my face into the winds of the future with something akin to panic.

I remembered arriving in the Indies on my first wedding anniversary, and now, nearly eight years later, I was going home alone in borrowed clothes. Widowed at twenty-six, with not a thing in the world that I could call my own, except my letters, a drawing and notebook, two spoons, a few trinkets, and a burned-out watch tied in Wiesje's handkerchief. In my mind's eye, I saw two lonely crosses on remote hillsides. Under the one lay the remains of my beloved Russell; under the other, the remains of dear Dr. Robert A. Jaffray. Suddenly I was awash in a sea of great bitterness. "Lord, I will never come back to these islands again. They have robbed me of everything that was most dear to me." I kissed Wiesje and thanked her for all she had meant to me through my years in Macassar. The young men had come back with the dinghy and pulled it onto the beach so that we wouldn't get our shoes wet; then they shoved it into the water. "I'll not look back!" But I heard the sound of running feet and voices calling, *Nona* Kemp! *Njonja* Diebler! *Selamat djalan!* "A peaceful journey!" Then the voices were raised in song, one I had sung many times with them at the close of services in the Tabernacle:

God be with you till we meet again!
By His counsels guide, uphold you,
With His sheep securely fold you;
God be with you till we meet again!

I had said I wouldn't look back but I did, and God broke my heart. Those were my friends from the Bible School, from the Tabernacle—Dyaks from Borneo, people from Sumatra, Bali, Lombok, Roti, Menado, Celebes, Aloer, and Ambon, all from the island world of the Indies, united as one in Christ Jesus, singing a farewell to us with tears running down their cheeks, hands waving. My own tears flowed down my cheeks, and all the bitterness was washed away. "Please, Father, forgive me. They are why I'm here! Not just because I was Russell's wife. I came because as a little girl You called me and I promised You I'd go anywhere, no matter what it cost! Forgive me, Lord!"

"Till we meet! Till we meet!
Till we meet at Jesus' feet;
Till we meet! Till we meet!
God be with you till we meet again!

Nona, Njonja, lekas kembali! "Miss, Mrs., come back quickly!" We waved and I called, "Someday I *will* come back again."

Strong arms helped us up the ladder, around the machine-gun mount, and into seats in the blister of the plane. They buckled us in with seatbelts. The engines were revving up, ready for take-off. I looked at my friends, still waving, and whispered, "Precious friends, someday I'll come home to you!"

CHAPTER TEN

From our seats in the blister of the plane we had a wonderful view of the rapidly receding coastline of Celebes. A smiling gentleman came through the door from the cabin, holding a first-aid kit in his hand. "I noticed your bandages when you came into the plane. Would you like me to put some clean dressings on for you?"

"Oh yes!" we both said.

A short while later, another smiling young man came bearing gifts—food in the form of sandwiches *made with white bread and butter!* We were on our second sandwich when the young man appeared with iced tea. He took the plate with the last sandwich on it off Margaret's lap so that she could hold her tea. She voiced what I had been thinking: "You won't throw that away, will you?" I had been wondering where we could put it, to save for later. It had been so long since we had seen bread and butter.

He appeared startled by her question; then he leaned over and said, "No, ma'am, I sure won't, and there's a lot more where this came from. So you just tell me when you'd like some more. Will you do that?" We assured him that we would and thanked him for being so kind. He beat a hasty retreat. I thought he looked as if he were going to cry, but he probably had hay fever.

A jeep was waiting when we landed in Balikpapan, Borneo, at the Australian army base. The driver greeted us with, "*Guh die!*" He spoke like the Australian major, so I knew that he was saying, "Good day!" He said that he would be taking us to sick bay in the nurses' quarters. We were given a towel and shown to the showers.

"Margaret, look, there's hot and cold water! O joy!"

"Yes, but have you smelled the soap?"

I really don't know how long we were in there—soaping, rinsing, and repeating the process. No pulling up buckets of cold water, and we didn't have to wash our clothes at the same time. What a wonderful invention, the shower! "Margaret," I called, "my hair squeaks; it must be really clean!"

Desperately weary, we sought our beds. I stretched out on the kapok mattress, thanking God for letting us be in such a clean place with showers and such caring people. How long I slept, I didn't know, but I awakened in a cold sweat; the bed was shaking and I was clutching the sides of it, thinking it was an earthquake and I was back in Barracks 8. If I fell, it would be a six-foot drop onto the dirt floor. My heart was pounding like a trip-hammer! Then I felt the iron bed frame and realized where I was. The mattress was resting on a metal frame with horizontal springs attached to it. I had turned in my sleep, causing the bed to shake. Margaret was having the same problem. She crept over to me and whispered, "Can you sleep? I keep thinking I'm falling out of bed. Do you think we could sleep on the floor?"

"That's a good idea. We could get back into bed before it's light."

"But Darlene, what if we don't wake up in time and they find us on the floor?"

"You're right. That could be very embarrassing. Maybe we'd better just stick it out." I wondered several times in the night why we felt our beds had to have springs! We both looked a bit haggard when the nurse came with a cup of steaming hot tea with milk and sugar and a small piece of wheat bread and butter. I felt greatly in need of that tea.

We dressed quickly. Word had come that they would be taking us on the first flight. Once airborne, a young army officer soon pointed out the large island that we were approaching, Palawan Island. The U.S. Navy had a large installation there, where we would be spending the night. They hadn't wished us to become overtired so had arranged our trip to Manila in three stages. The navy was expecting us and knew who we were: the last of the American women POWs to be evacuated.

When the Catalina came to a stop on the runway, we were received by navy personnel in gleaming white uniforms. The jeep stopped on the road by a cement sidewalk leading up to a white building. It came to me that they were not very well camouflaged; just as quickly, I reminded myself that the war was over and they didn't have to worry about a bombing raid. Men lined either side of the walkway. It looked like an honor guard, and it worried me that they didn't know we were just ordinary people. The officers guided us up the steps with a hand under our elbows. As we stepped into the large mess hall, a band began to play "The Star-Spangled Banner," and Old Glory was unfurled before us. I came undone. I couldn't help it; I began to sob. This was the first American flag we had seen in years. No one who had not experienced the past four years with us could understand what it meant to see that flag and hear that song. I thought my heart would burst within me with pride and with the first-born feeling of being really free! These are my countrymen. "O God, I'm free! I may still be classed as a POW, but I'm free."

An arm went round my shoulder and someone was saying, "There, there, don't cry. It's all right now, and you're safe here!"

They all looked immaculately groomed, and there was a dignity and an unaffected aura of self-esteem that emanated from them. I could not but contrast them with the men of Pare Pare, with their broken, bowed, cadaverous bodies (even though some had said they had gained twenty pounds). In Pare Pare the clothes were mismatched and ill-fitting. The men were either barefoot or in thongs. I didn't remember seeing any shoes—certainly nothing to be compared with the highly polished black footwear I had seen here—and yet there was a dignity and a relaxed poise of self-knowledge evident in the men of Pare Pare too, qualities of character forged in the fires of affliction while learning to endure!

Was it right to contrast them? Perhaps these young men were among those who had landed on this island and gone through the hell of bombings, snipers' bullets, grenades, machine gun fire, and hand-to-hand combat—the terrors that had taken the lives of their buddies, who lay beneath the countless white crosses in the large cemetery I had seen from the plane. Then these men, too, had *earned* their badges of courage, their dignity, and their poise from battles fought and victories won while learning to endure. I felt pride and pleasure in them all—my Dutch friends and my fellow Americans.

Many kind men waved us off the next morning as we left on the last leg of the journey to Manila by plane. We found a bus waiting when we deplaned. We had a good driver, one who would have qualified for the Indianapolis 500! First on the horn had the right of way. But we arrived safely at the POW camp and were ushered into a long building with beds on either side. The showers and mess were pointed out, and we were told that lunch would be served in a half hour.

While we were eating, a doctor came to ask our names and where we had come from. We were scheduled for thorough physicals that afternoon. In the meantime he wanted us to take an enormous number of vitamins and minerals, and since I had had malaria, I had to start on Atabrine immediately. "It will make you very yellow, but it's an improvement over quinine."

I told him that I knew he didn't want us to eat much; after taking all those vitamins and minerals before each meal, there would be no room for food. He laughed. "No, I want you to eat well. You've had a rough time, haven't you? I want you to get a lot of rest while you're here." As part of the cure, he made appointments for us both to have permanents: "Good for the morale!"

That evening we met Mary and Herman Dixon, our fellow missionaries from Borneo. They had survived the camp in Kuching, Sarawak.

Daily for more than a week, I haunted the POW camp post office asking for mail for Darlene Deibler or Margaret Kemp. One day the young mail clerk leaned across the counter and said, "Boy! I don't know why someone wouldn't write to you!" I was so embarrassed that I decided not to go back again until I learned we were leaving for home. I didn't have any explanation to give him as to why no one was writing to me; I didn't understand it either.

After several weeks there was an announcement that a Dutch ship, the *Klipfontein*, on loan to the Americans for evacuating army personnel and POWs, would be in Manila in a few days. It would take on fresh stores and water; then it would begin boarding passengers. Lists of the passengers were posted; our names and the Dixons' were there, with those of other POWs.

A bus took us to the pier. As Margaret and I stood looking at the mass of people waiting to board, we decided that there was no way all those adults and children could be accommodated on that ship, but they all got on board.

We single ladies were accommodated on deck, on the stern of the ship, three deep in hammocks under a canvas cover. There was plenty of fresh air, but little privacy and less space.

The lighting under the canvas wasn't good, so we usually took our Bibles and sat on the deck in the open. Many sat down next to us and shared their problems. I was not on this ship by chance. There were those who had received Dear John letters; others had had letters from their wives asking for a divorce. Some were ill. The twenty-three-day trip from Manila to San Francisco passed quite quickly.

The closer we came to San Francisco, the more excited the passengers became. The afternoon we saw the Golden Gate Bridge in the distance, shining like gold in the last rays of the setting sun, a cheer went up that was deafening. Then someone broke out with "I Left My Heart in San Francisco"! The more they sang and the more they extolled the beauty and the wonders of this fabulous city, the more alone and isolated I felt. I began to panic. When we disembarked, I would still be half a continent away from home. "Margaret, do you know anyone in San Francisco?" She didn't, and looked as lost as I felt. "Well, we'll just sit on the dock on our rattan cases until the Red Cross shows up!"

"Aren't you excited?" several asked. "You'll just love San Francisco!"

"But I don't know anyone here, and I need to get to Iowa to trace my family. I *do* have an aunt and uncle in Ontario, California, but I've lost their address."

"Ontario? That's way down in southern California, a long way from here. But don't worry, the Red Cross will help you."

Just then an announcement came over the loudspeaker that turned the cheers to groans. "We have just been informed by the Port Authority that the harbor is full. There are no empty berths, nor will there be for days. We have been instructed to continue on to Seattle, Washington." So we turned and started north. I knew one person who was happy for the change. I was greatly relieved that the Lord had given me a few more days to work out my problem of what to do and where to go when I got off. I had never traveled alone, much less without identification papers or an address book or some money.

It was afternoon when we docked. I began to feel tremulous inside. Again there was an announcement on the loudspeaker: it was Navy Day in Seattle, so they would begin processing us in the morning. "O joy," I thought. "I'll have a few more hours on known territory and with friends." The Dixons said we could go with them, as Mary had a brother in Seattle, so one problem was taken care of.

The following morning, when people began to disembark, Margaret came to say that the Meltzers, missionaries on furlough from Borneo, were there and had asked her to go with them. I waved and watched Margaret leave, thinking how much I would miss her. The Dixons were nowhere in sight; someone said he had seen them leave to go shopping and they would be back. Feeling deserted, I leaned on the rail, looking for a familiar face, when suddenly it struck me: Dad and Mother are gone, that's why I haven't heard from them! I ran under the canvas, got down under the hammocks, and wept in great agony of soul: "O God, You took Russell; did You have to take Mother and Dad also? Now I have no one!" I cried and cried, feeling totally abandoned. Then He came and I knew my Lord was there: "My child, you can still trust Me. I told you that I would never leave you nor forsake you."

With the sound of His voice, speaking deep within me, a great calm settled over me. I dried my tears. "All right, Lord, it's You and I against that strange world out there!" I crawled out from under the hammock, straightened my coat, and went in search of a Red Cross worker. I rounded the corner of the deck and there she was! I latched onto her. "Please, you have to help me. I'm a POW and all my papers were burned when we were bombed. I need money to get to Boone, Iowa, to trace my family, if any of them are still alive."

"What's your name?"

"I'm Darlene Deibler."

"Honey, I've been on this ship all morning looking for you. I have three telegrams here, and they're all from your mother and dad."

My hands were trembling so violently that I couldn't open them, so she did it for me. I read, "Welcome home. We've been in contact with the Red Cross and knew you were a passenger on the *Klipfontein*. We were going to meet you in San Francisco, but when they diverted your ship to Seattle, we knew we couldn't get there in time to meet you, so have sent money for you by Western Union so you can come by train to Oakland. We moved to Oakland in 1942. Call us collect as soon as you get to a phone. Love, Dad, Mother."

I began watching people on the pier and noticed that the ladies were all wearing coats with very short nap. Looking down at my long-haired coat, I felt like a big, shaggy bear—really tacky! I made an instant decision and went below to find the captain. When I asked to borrow his razor, he looked puzzled but asked no questions. "Of course!" he said and went to fetch it from his cabin. Armed with his razor, on a remote corner of the deck, I gave my coat the new short-nap look. It really was quite smart-looking after its "close shave"! I returned the razor, thanked him for the pleasant trip, and said goodbye.

First stop for me was the Western Union office, then the depot to purchase my train ticket. I had to be sure I had enough money for that and food before I bought anything else. When I stepped up to the window and asked for a one-way coach ticket to Oakland, California, I didn't expect the response I got from the clerk. He slapped his hand against his forehead as though in shock. "My dear! Don't you know that a war has been on and only army and navy personnel can get reservations?"

If the despair I felt showed in my face, I don't know, but I felt sick. "No, sir," I explained, "I didn't know that. I just came in on a ship. I've been a prisoner of war for almost four years. I just found out that my mother and father are living in Oakland, California, and I'm trying to get there." I was close to tears.

He reached behind him, grabbed a ticket, and said, "There, there! Now don't cry. I've got lots of tickets for people like you!"

I was in business again; the panic gave way to relief. "Oh, thank you. I didn't know what I was going to do."

"Now don't you worry. You be here at seven. The porter will help you with your bags, and he'll show you which coach to get on. Your seat number is on the ticket, and if you need anything, just ask the conductor. You should be in Oakland tomorrow at eleven. Have a nice trip!"

I thanked the clerk again, then sat down in the waiting room and tied my ticket and money into my handkerchief and put it safely into my dress pocket. With my coat securely belted, I went in search of a telephone. When I got in the booth and looked at the contraption in front of me, I couldn't believe what they had done to telephones. There was a circular metal affair with letters as well as numbers around it. I thought, "How do you work this thing?" I didn't take time to read the instructions, I just panicked.

A gentleman waiting outside one of the other booths asked if something was wrong and if he could help. I explained my dilemma: I had been out of the country for eight years, my parents had asked me to call collect, and I didn't know how to use the phone. I had never seen one like this.

"Do you know the number?" I showed him the number in the telegram and he offered to put through the call for me. He needed to know my name also, he said, as the operator would ask. I told him and presently he handed the receiver to me, saying, "She's ringing now."

I was shaking like a leaf; this was a big moment and my heart was pounding. I heard the receiver go up on the other end and a voice said, "Hello, Darlene."

"Hello, Mother." I couldn't say another word, I was crying so hard.

Mother was so calm and reassuring. She told me that my brother, Raymond, had just gotten back to the States from Germany and that our sister, Helen, and her family were there to meet me in Oakland. My older brother, Donald, and family were fine. Every time she paused, I said, "Uh huh." That's all I *could* say. I think she covered most of the family history before saying, "All right, I'll say goodbye for now. Get your ticket and we'll meet you here in Oakland. Goodbye." To which I made the intelligent reply, "Uh huh."

I stood there until I had stopped crying and then went through the mopping-up process with the corners of my handkerchief-cum-purse. How like the Lord Mother was; she understood that all I could say was, "Uh huh," but I had needed so much to hear about the family, and to know that they had sent many letters through the years—all of them returned marked "Missing person, no trace of her." Mother had saved them to show me when I got there. How many times had I come to the Lord in these past years, in such agony of grief, fear, pain, and loneliness, and all I could say was, "Jesus." How tenderly He talked to me, reminding me of promises He had made to me, counseling and sometimes rebuking, when all I could say was, "Uh huh, dear Lord Jesus."

Many have asked me how I know it is the Lord speaking to me. What had just happened was the best illustration I know. I hadn't heard my mother's voice for over eight years, but when the receiver went up in Oakland, California, and I heard someone say, "Hello, Darlene," I knew it was Mother. No one ever spoke my name as she did. So it is, that when I hear deep within the recesses of my spirit Someone say, "My child," I know it is my Lord. No one else calls me as He does. That is His promise to all His children in John 10: The sheep hear His voice and He calls His own sheep by name, and leads them out. The sheep follow Him: because they know His voice! "I am the good shepherd; and know my sheep and am known of mine."

The Dixons joined me at the depot. The train trip was wonderful. The first thing I noticed was how much the cars had been updated. The leatherette seats were no more; they were now covered with what looked like velvet. There were clean white covers on the headrests. The tables in the dining car had freshly starched white tablecloths and napkins. There were flowers in silver vases on each table. Everything was so clean—no open windows to let in the dust.

Back in the coach, settling down with a pillow (in a sparkling white, ironed pillowcase) behind my head, I felt extremely weary. What a day! I thought of the kind people God seemed to have everywhere to help me when I needed it, and the phone call from Mother—but how could I have forgotten? I hadn't told Mother what time the train was arriving in Oakland! (In fact, I didn't tell her much of anything but "Uh huh.") Herman said I could send a night telegram from Portland, Oregon. The train made a half-hour stop there. My parents would have the word in plenty of time, because the telegram would be phoned through immediately. I settled on "Arriving at eleven this morning stop kill the fatted calf love Darlene." It was two o'clock in the morning when we pulled into Portland. We went immediately to send our telegrams, then took a walk in the crisp, cold air of fall.

We breakfasted in the dining car, then returned to check our bags and watch the changing scenery. I had never seen such magnificent trees as the redwoods of California. When the train tracks ran parallel with the highway, I noticed the cars. I was amazed at the change in the body lines. They were long and sleek, with what looked like fins on the back fenders. Eight years made quite a difference. It was like a Rip Van Winkle story—the amazing advances in technology, no doubt accelerated by the war, and the changes in every sphere of life, with everyone in a hurry. I had heard of television, for example, but had never seen it.

I glanced across the aisle at a smiling GI and his family—what a beautiful sound, the "happy talk" and the laughing, well-fed, beautifully dressed children. "Lord, will I ever be young and carefree again? What am I now—sixty going on a hundred? I feel strange among all these sophisticated people." With blessed assurance, the verse came to mind that the Lord had given me at the beginning of the war; Deuteronomy 33:12, "The beloved of the Lord shall dwell in safety by him; and the Lord shall cover (overshadow) him all the day long, and he shall dwell between his shoulders." What a wonderful, reassuring verse when I was walking into the unknown, with only my family as a point of reference.

In the morning the porter came through the car collecting pillows, and the conductor collected our seat reservation tickets. "Oakland! Next stop Oakland!" As the train slowed I strained to see the faces of those so dear to my heart. Would I recognize them? Would they know me? I came down the steps with my suitcase, looking into the sea of faces, searching for . . . then I saw a hand go up and heard, "Darlene! Darlene!"—another voice instantly familiar, my daddy's! Looking back, he called, "Mother, she's here!" Then they came running, my precious father, more handsome than ever, and my dear mother, now calling my name. I ran to meet them, holding and kissing them, not wanting to let them go. We were crying as I whispered, "Oh, Mother, Daddy, I thought so many times that I would never see you again!"

They kept saying, "It's all right now, honey, you're home. You're safe." Then my precious sister, Helen, her husband, Clarence, and their daughter, Coralyn, caught up with us and our arms opened to encircle them. What a joyous homecoming! How dear these ones were to me; they were my family and they were my friends. The long years of separation left no reserve or strangeness; they were as my heart remembered them, unchanged in a world that had changed so much. Coralyn was two when I saw her last, and now she was ten; but I recognized her beautiful dimples when she smiled. There were a number of people from the church my parents attended also there to welcome me.

My father told me about the drama of the 2:00 A.M. telegram. The household was awakened when the phone started ringing. Mother dashed out to answer it. All dad could hear after the first hello was, "What?" There was another long pause, to which mother answered, "What was that? Would you please read that again?" Then a very long pause.

The Western Union operator evidently knew the story of the Prodigal Son, for she said, giggling, "Sounds like a wanderer is coming home."

Then my father heard mother say, "Oh, yes, of course. Thank you!" She ran into the bedroom crying and grabbed my father, shaking him and saying, "She's all right, Orvis. She's all right; she hasn't lost her sense of humor!" Then she told him what the telegram had said. I don't believe my "Uh huhs" had been too reassuring! By then the household was awake. That called for a cup of coffee and a discussion, after which they all settled down for a short morning's nap.

We had so much visiting and catching up to do that it was very late before we went to bed. Long after the others slept, I lay awake, thinking about my arrival in Oakland and the inexpressible joy that I knew in seeing my loved ones again. The thought that came to me when I remembered looking over their heads at the beautiful blue sky dotted with fleecy white clouds was this: "If it's this joyous to see my loved ones, who have walked with me by prayer along the trails of New Guinea and through the streets of Macassar; these dear ones, who suffered and prayed for me during the war years, when the newspapers carried the horror stories of the atrocities perpetrated by the shock troops and the Kempeitai; if seeing them is so overwhelming, what will it be like when I see Jesus, my beloved Lord, Who walked the same trails and streets with me and Who never left my side during those long years of suffering and sorrow? What will it be like?"

Viewing those eight years from this far side, I marvel at the wisdom and love of our God, Who controls the curtains of the stage on which the drama of our lives is played; His hand draws aside the curtains of events only far enough for us to view one sequence at a time.

Had those eight years been revealed to me in one panoramic view that misty gray January morning in 1938, would I have had the courage to board the ship? I wonder.

Through the intervening years, tempestuous winds of gale force have buffeted me. Waves of tidal proportions have threatened to carry me under or dash me upon the rocks. But knowing now what I did not know those many years ago, with C. H. Spurgeon, I can thank my God for every storm that has wrecked me upon the Rock, Christ Jesus!

EPILOGUE

Darlene Deibler arrived in Oakland, emaciated and emotionally fatigued, to be welcomed into her family's love and care on November 30, 1945. The twenty-three pounds that had been starved from her body returned slowly, as did her physical and emotional reserves. Over the next two years, Darlene testified to the power and presence of God throughout her prison experiences before many who marveled at the fact that she had survived at all.

Time eased her grief over Russell's death, while her memories of their life together in New Guinea confirmed her calling and the necessity to return. She had been called to serve as a missionary long before she met Russell. She resisted the many words of advice against single women missionaries, especially one as young as she, as well as the encouragement to stay home and let some years of comfort repay her pain in the South Pacific.

In 1946 a young man, Gerald W. Rose, was given a film to use in deputation. It was a documentary of Rev. C. Russell Deibler's trek to the Wissel Lakes in the interior of Dutch New Guinea. Rev. Gerald Rose was already under appointment to this primitive mission field. Mutual friends arranged a meeting between Darlene and Jerry, unbeknownst to either of them. As it was in God's plan, Jerry and Darlene married on April 4, 1948, and they began their ministry in New Guinea in early 1949.

Together Darlene and Jerry returned to the Wissel Lakes area and later pioneered the work among the Dani tribe in the Baliem Valley and the lower Wahgi Valley of Papua New Guinea. Their two sons, Bruce Gareth and Brian Jaffray, were raised among the native peoples.

Darlene and Jerry stayed in New Guinea until 1978, when their station, Nondugl, was expropriated by the newly independent nation of Papua New Guinea. They then moved to meet yet another challenge in the Australian Outback. Darlene and Jerry presently live on a station five hours drive south of Darwin, in the Northern Territory near the village of Larrimah.

Darlene left Celebes in 1945 knowing that Mr. Yamaji had been sentenced to be executed for killing the man at Pare Pare; but because of his kindness to Darlene while she was in the Kempeitai prison, his sentence was commuted to life imprisonment with hard labor. Still later that sentence was also commuted.

In 1986 Darlene and Jerry visited Elsie David, a fellow Kampili internee living in Australia. She related the following story: A friend vacationing in Java happened upon a priest who had just returned from bicycling in Japan. While in a small coastal village, the priest had stopped at a bicycle shop for repairs. Striking up a conversation with the owner, who spoke Indonesian, the priest discovered that the man had been the commander of the women's POW camp outside of Macassar during World War II. The owner asked the priest, if he ever met any of the women who had been in Kampili, to tell them he was sorry he'd been so cruel. He said he was a different man now. Darlene felt the remark affirmed that Yamaji had indeed had a lasting change of heart.